CLEARING THE WAY

CLEARING THE WAY

U.S. Army Engineers in World War II

CHRIS MCNAB

CASEMATE

Philadelphia & Oxford

Published in the United States of America and Great Britain in 2023 by
CASEMATE PUBLISHERS
1950 Lawrence Road, Havertown, PA 19083, USA
and
The Old Music Hall, 106–108 Cowley Road, Oxford OX4 1JE, UK

Hardback Edition: ISBN 978-1-63624-386-3
Digital Edition: ISBN 978-1-63624-387-0

A CIP record for this book is available from the British Library

Printed and bound in the United Kingdom by CPI Group (UK) Ltd, Croydon, CR0 4YY
Typeset in India by DiTech Publishing Services

For a complete list of Casemate titles, please contact:

CASEMATE PUBLISHERS (US)
Telephone (610) 853-9131
Fax (610) 853-9146
Email: casemate@casematepublishers.com
www.casematepublishers.com

CASEMATE PUBLISHERS (UK)
Telephone (0)1226 734350
Email: casemate-uk@casematepublishers.co.uk
www.casematepublishers.co.uk

Front cover image: Engineers of the 167th Engineer Combat Battalion, 1117th Engineer Group,
construct the first Bailey bridge across the Rhine River in the vicinity of Wesel, Germany.
They worked continuously for 12 hours under enemy fire. (NARA, ETO-HQ-45-23747.
Photographer: Drummond)

Back cover image: Infantry footbridge supported by pontons across the Roer river in Germany,
February, 1945. (U.S. Army)

Contents

Introduction

In the case of the United States Army Corps of Engineers (USACE) in World War II, we can make the boldest of claims—without them, the U.S. armed forces would not have been able to fight the war. To put some meat on this argument, a sound place to start is a quotation from the War Department's *Engineer Field Manual: Volume I Engineer Troops* (1932), which opens by clarifying the mission and functions of the engineers:

> 1. Mission.—The mission of engineers in war is to assist the operations of the field forces by means of engineering works.
> 2. Functions.—*a.* The functions of the engineers in the theater of operations include—
> (1) All work of construction and the repair and maintenance of all structures of every character, except such as are specifically assigned to other arms and services.
> (2) Military mining, demolitions, and protective measures against enemy mines.
> (3) The operation of railways, portable and fixed electric light and power systems, water supply systems, and all other utilities of general service, except such as are specifically assigned to other arms and services.
> (4) The execution of surveying and mapping, including the production and distribution of maps.
> (5) The procurement, storage, and issue of all materials for construction work, for the organization of defense systems, and for all other operations assigned to the engineer arm, including all plant, tools, and appliances for such work.
> *b.* The most important function of the Corps of Engineers is the maintenance, improvement, and construction of routes of communication and movement. This is a continuous operation and employs the major portion of the engineer personnel.

What is immediately apparent here is the sheer breadth of the engineers' duties and also the magnified importance of their responsibilities. The USACE, often in a literal sense, built the foundations for the functionality of the U.S. Army. Without the engineers, U.S. combat aircraft would have had few forward airfields from which to operate. Without the engineers, U.S. amphibious forces would have been unable to land on contested beaches in North Africa, Sicily, Italy, Normandy, and across the Pacific. Without the engineers, millions of tons of supplies would have been stranded out at sea, their shipping lacking viable ports, anchorages, and storage depots. In the combat theaters, infantry and armor advances were often only made possible by the fact that engineers had built or restored the communications infrastructure ahead of them—roads, bridges, railroads. Engineers ensured that millions of soldiers had sanitation, hospitals, camps, showers, barracks, light, heating, and much, much more. Even the U.S. atomic bomb program, the Manhattan Project, would not have been possible without the engineers' efforts in constructing plants, facilities, and vital equipment. Taken together, the efforts of the USACE, including their civil works, ensured that the U.S. Army emerged victorious in World War II. Indeed, one could make plausible arguments that the emergence of the United States as the world's greatest superpower was heavily indebted to U.S. military engineering.

By the beginning of World War II, the U.S. Army engineers already had a venerable history. Engineer officers became part of the emerging U.S. Army during the American Revolutionary War (1775–83)—the first engineer officers were appointed by George Washington on June 16, 1775—and in 1802 the Corps of Engineers were established as a separate branch of the armed forces. On July 4, 1838, a further engineering for-mation was founded, the U.S. Army Corps of Topographical Engineers, responsible for mapping, defining navigational routes, and for designing and building federal civil works, although this corps was eventually folded into the larger Corps of Engineers in 1863.

Compared to the vast entity they would later become, the engineers during their first century of existence were comparatively diminutive. But they nevertheless demonstrated impressive utility. In terms of civil works, the engineers laid down many of the transportation arteries—roads, bridges, canals, railroads, ports—through which the blood of American commerce and settlement flowed with gathering strength. On the military side of their activities, they were heavily

A U.S. Army recruitment poster glamorizes the darker reality of conducting engineering work under fire. (NARA)

employed in the design and construction of concrete coastal artillery fortifications, many of which remain standing to this day. The engineers also demonstrated their invaluable applications in active combat theaters. In the Mexican–American War of 1846–48, for example, engineers under the leadership of Chief Engineer Joseph Totten played a pivotal role in the siege of the port of Veracruz in March 1847. During the national convulsion of the American Civil War (1861–65), engineers were at the forefront of campaigns on both Confederate and Union sides, pioneering new combat engineering skills, especially in the disciplines of erecting field fortifications and making river crossings. In December 1862, Union engineers famously constructed, under Confederate fire, a floating ponton bridge across the Rappahannock River at the battle of Fredericksburg, the first example in U.S. military history of American troops making a combat water crossing and establishing a beachhead on the other side. It is no coincidence that some of the most famous generals of the Civil War came from an engineering background, including, on the Union side, George McClellan, Henry Halleck, and George Meade, and Robert E. Lee, Joseph Johnston, and P. G. T. Beauregard for the Confederates.

Following the Civil War, the Corps of Engineers kept militarily active through the assorted expeditionary conflicts in which the United States became embroiled. These included the Spanish–American War of 1898, where deployments to Cuba, Puerto Rico, the Philippines, and Guam showed that the engineers could adapt to tropical theaters. (The experience would later prove very useful in the 20th century.) Official recognition of their utility was evidenced in their steady growth—by 1901 the Corps was three battalions strong, of four companies each. The reputation of the engineers for getting the job done was also greatly bolstered by the USACE's central involvement in the construction of the Panama Canal between 1904 and 1914. After the failure of two civilian chief engineers to jump-start the project, President Theodore Roosevelt appointed Major George W. Goethals of the USACE to the role. (Goethals had the added productivity incentive that if he suddenly walked off the job, he would be court-martialed.) He was an inspired and supremely effective choice. Hewing the 50-mile canal, which included building two dams, six sets of locks, two artificial lakes, a regulating works, a telephone and telegraph system, a hydroelectric station, and a railroad, remains one of the greatest civil engineering projects in history.

As with many other branches of the U.S. armed forces, it was World War I (1914–18) that transformed the USACE in both scale and scope. Although the United States only entered the war in 1917–18, by the end of the conflict some 240,000 engineers had served in Europe, with each U.S. Army combat division including an organic 1,660-man engineering regiment. The engineers made possible the deployment and maintenance of the American Expeditionary Force (AEF), especially in creating the communications and transport infrastructure feeding the frontlines. Feats performed by the engineers on the Western Front include producing about 200 million feet of lumber, laying down 950 miles of standard-gauge rail lines (in addition to the hundreds of miles of narrow-gauge lines), building 20 million square feet of storage space, and erecting barracks sufficient to hold 742,000 men. In action, the engineers also proved their capabilities as skilled combatants, applying their technical skills to cutting barbed wire, demolishing strongpoints, and establishing defensive positions under fire. Participation in combat explains some of the high casualty rates suffered by engineer units; the 2d Engineer Regiment, 2d Infantry Division, for example, had a 12.73 percent casualty rate.

Post-World War I, the USACE underwent the scything demobilization experienced by the wider U.S. forces, dropping to just eight or nine regiments on active service, most of those employed in civil works projects. But within two decades, a new world war was on the horizon, the flames beginning in Europe in 1939 but eventually spreading to encompass much of the planet. From 1940, the USACE therefore began a period of inexorable growth, the trajectory of which steepened sharply once the United States was plunged into the shooting war from December 7, 1941, with the Japanese attack on Pearl Harbor. By June 1945, the USACE had mobilized 89 divisional combat battalions, 204 nondivisional combat battalions, 124 aviation battalions, 79 general service regiments, and 36 construction battalions, plus hundreds of specialist companies, with more than 750,000 men serving in the engineers. Notably, during the war Black Americans came to constitute about 20 percent of USACE personnel, although most were used in labor roles.

It is impossible to summarize briskly the achievements and contributions of the USACE in World War II, although we will explore some of their key feats and actions in the introductions to the content of subsequent chapters. Select any U.S. military campaign of the war, in any theater, and even the most cursory research quickly reveals the absolutely central influence of the engineers to the path and outcome of the fighting. What is particularly striking is the encyclopedic knowledge of engineering required within the Corps. To demonstrate, we again return to the pre-war *Engineer Troops* manual. The following passage, quoted at length to make the point, describes the responsibilities of engineers in an active theater of operations. Note that this list excludes the equally exhaustive activities expected of the engineers performing work in the "zone of the interior," i.e., the continental United States:

> Engineers in the theater of operations include among other activities the construction, repair, and maintenance of roads and trails; of bridges and other means for crossing rivers and similar obstacles; of shelter for troops and animals, including huts, hospital buildings, barracks, stables, and accessory structures; of storehouses, shop structures, hangars and flying fields, including, in proper cases, the installation of the necessary machinery; of wharves, railroads, and light railroads; the provision of water supply, including sterilization in bulk; the provision and installation of baths, disinfectors, dipping vats, and incinerators; the installation of plumbing, sewage disposal, and heating plants; the installation of machinery for refrigerating plants, laundries, and other mechanical plants; assistance to other arms in intrenching, and in organizing defensive lines; the organization and construction of rear lines of defense; the construction of bombproofs, observation stations, machine-gun emplacements, and other special works of defense; the execution of special measures for destroying or overcoming enemy obstacles; the supply of camouflage material and the supervision and inspection of its use; the preparation of signs for the direction of troops, including road signs, traffic signs, signs indicating the location of water points and other establishments, and signs safeguarding against the use of impure water; the operation of electric light, gas, and power plants and water supply plants; the operation of shops for the erection and repair of railroad rolling stock, of construction machinery of all kinds, and for the manufacture of special appliances for engineer operations; photographic and cinematographic work pertaining to terrestrial reconnaissance, and terrestrial surveying; map making, map reproduction, and map supply; the training of engineer troops for all their duties, and the compilation of technical data and the preparation of training literature on subjects pertaining to any of the operations assigned to the Corps of Engineers.

The U.S. Army engineers of World War II performed all such duties, and also tackled the thousands of other unforeseen and random challenges thrown up by a conflict across every conceivable type of terrain. This book attempts to shed light on how the engineers performed just a fraction of these roles, through a selection of texts from official engineering manuals of the period. A word of warning here. Compiling this book is an exercise in being painfully selective. The total library of engineering manuals from the 1930s and 1940s runs to millions of words across thousands of topics, so here we can only scratch the surface.

Yet the collection does reveal those elements that are crucial to understanding the roles and capabilities of the USACE soldiers. First is their breadth of technical expertise. Essentially, an engineer regiment contained within it the talent sufficient to plan, design, build, and run

Men of the 65th Engineers hard at work constructing a machine-gun pit on Hawaii in 1942. (Signal Corps Archive)

a small town, on top of which came all the hard engineering skills specific to warfighting, from designing a fortification to blowing one up. Second, adaptability was surely the key characteristic of the engineers. War zones are by their very nature defined by emergency, improvisation, unpredictability. In this environment, the U.S. Army engineers couldn't always do things "by the book"; they had to work with whatever they had wherever they were, often with the enemy actively striving to stop them. Uncomplaining innovation was, and remains, central to the engineering personality. Finally, engineers had to be tough. The pressures on them to perform tasks that, in civil contexts, would have been regarded as impossible were simply immense. In frontline engineering, sleep became an extravagant and rare luxury, and the penalties for failure were way beyond a stern ticking-off by management—the fate of entire campaigns and thousands of personnel could hang on engineering efficiency. Thousands of combat engineers actually fought shoulder to shoulder with the infantry, and the USACE took more than 29,000 casualties during the World War II. Here was engineering truly at the extremes of both technical skill and human endurance.

CHAPTER I

Engineer Roles and Units

Because of the vast spectrum of roles assigned to the USACE in World War II, both combat duties and civil works, it can be more useful to think of the formation in terms of what historian Martin Reuss has called a "major international organization," one which "combined the functions of a school, research laboratory, department store, shipper, engineering firm, repair shop, and construction organization" (Fowle 1992: 3). Dividing the Corps of Engineers along the crudest lines, we can separate its functions into those performed by its military section and those by the civil works section. The former was tasked with direct engineering support to military units, while the latter focused on providing critical work stateside, such as ensuring key waterways remained open and navigable, and building and maintaining hydroelectric power plants.

All aspects of the USACE fell under the authority of the Headquarters of the Corps of Engineers, led by the Chief of Engineers, a role with one of the broader and more complex task portfolios in the U.S. armed services. In November 1941, on the eve of American entry into World War II, the Office of the Chief of Engineers presided over four main divisions, each headed by an assistant chief: Construction Division, Supply Division, Troops Division, and Administration Division, plus a collection of boards and commissions with specialist responsibilities, such as the Mississippi River Commission and the Beach Erosion Board. A separate Engineering Division was established in May 1943. More radical changes came in late 1943 and early 1944, as the five engineering divisions multiplied to reflect some of the decentralized demand and command realities of a global war. Nine divisions were established in November 1943—Procurement, Supply, International, Engineering and Development, Military Intelligence, War Plans, Civil Works, Military Construction, and Real Estate—while a further two divisions, Maintenance and Readjustment, were added in the first half of 1944. Further modifications came in April 1945, as the engineering responsibilities increasingly came to reflect the predicted end of the war.

The following manual extract comes from the War Department's Engineer Field Manual: Engineer Troops *of 1943. It usefully describes how engineer units were organized and assigned down to company level. It is important to note that an engineer officer (known as the "unit engineer") would be placed within the staff of every command to which the engineer units were assigned—from the top down, engineers were involved in operational planning and execution.*

A mess hall built from scratch by U.S. Army engineers in 1942. (LOC)

From FM 5-5, *Engineer Field Manual: Engineer Troops* (1943)

CHAPTER 1
GENERAL

■ 1. PURPOSE AND SCOPE.—This manual is designed to serve as a general reference on engineer organization. It covers the mission, classification, organization, equipment, armament, and training of engineer units.

■ 2. CLASSIFICATION.— Engineer troops are classified as combat units or service units in accordance with Circular No. 422, War Department, 1942.

a. Combat classification.—(1) Combat units are those whose functions require close contact with the enemy. They include general engineer and some special engineer units attached or assigned to divisions, corps, and armies. They receive extensive combat and tactical training.

(2) Engineer headquarters of corps and higher units, including all bases and defense commands, are in the combat classification.

b. Service classification.—Service units are those whose functions are characterized by service activities rather than combat. They include some general engineer and a majority of special engineer units attached to corps and higher headquarters. They are organized, trained, and equipped to do the more technical and permanent engineer work. Service units receive less extensive combat and tactical training than combat units.

■ 3. ASSIGNMENT.—The Army of the United States is divided into Army Ground Forces, Army Air Forces, and Army Service Forces. Regardless of where or with what units they are serving, engineer troops generally are identified with one of these three forces, as shown in the following paragraphs.

■ 4. ENGINEER UNITS, COMBAT, WITH GROUND FORCES.—*a. Engineer combat battalion.*—This battalion is an organic part of the infantry division. It consists of a headquarters, headquarters and service company, three lettered companies, and a medical detachment. It is completely motorized, including transportation for all personnel.

b. Engineer combat battalion, nondivisional (attached to corps or army).—This unit is organized, trained, and equipped the same as the engineer combat battalion of the infantry division. The number of battalions attached to corps and army depends upon the situation.

c. Engineer squadron. This squadron is an organic part of the cavalry division. It consists of a headquarters, headquarters and service troop, two lettered troops, and a medical detachment. It is completely motorized, including transportation for all personnel.

d. Armored engineer battalion.—This battalion is an organic part of the armored division. It consists of a headquarters, headquarters company, four lettered companies, and a medical detachment. It is completely motorized.

e. Engineer motorized battalion.—This battalion is an organic part of the motorized division. It consists of a headquarters, headquarters and service company, three lettered companies, a reconnaissance company, and a medical detachment. It is completely motorized, including transportation for all personnel.

f. Engineer mountain battalion.—This battalion is an organic part of the mountain division. It consists of a headquarters, headquarters and service company, a motorized company, two pack companies, and a medical detachment. It has not sufficient transportation to move all personnel and equipment,

g. Airborne engineer battalion.—This battalion is an organic part of the airborne division. It consists of a headquarters, headquarters and service company, a parachute company, two glider companies, and a medical detachment. Personnel, armament, vehicles, and equipment required for each mission are transported by aircraft to the scene of operations.

h. Engineer light ponton company.—This company is equipped with the M3 pneumatic bridge, which can handle all normal infantry division loads and may be reinforced to carry heavier loads. It has two bridge platoons each equipped with one unit of M3 pneumatic bridge, and a light equipage platoon which has one unit of footbridge and equipment for ferrying. The company is an organic unit of army and higher echelons *i. Engineer heavy ponton battalion.*—This battalion is equipped with heavy ponton equipage to provide means of stream crossing for military vehicles too heavy to pass over a light ponton bridge. It has two lettered companies of two bridge platoons each. Each bridge platoon is equipped with one unit of heavy ponton equipage. The battalion is an organic unit of army and higher echelons.

j. Engineer treadway bridge company.—This company consists of company headquarters and two bridge platoons. It is an organic unit of the armored force, and normally is attached to an armored engineer battalion. Each bridge platoon transports one unit of steel treadway bridge equipage for construction of ferries and bridges in river-crossing operations of the armored division.

■ 5. ENGINEER UNITS, SERVICE, WITH GROUND FORCES.—*a. Engineer light equipment company.*—This unit consists of a company headquarters and two equipment platoons. It furnishes supplementary equipment, with operators, to combat battalions, and also operates as a replacement pool for construction equipment. It is attached to corps or army.

b. Engineer depot company (with Army Service Forces as swell as Army Ground Forces).—This company operates engineer depots and other engineer supply points. It has three depot platoons, and a depot section in the headquarters platoon. Organic equipment is that necessary for operating depots. It is attached to army and higher echelons.

c. Engineer parts supply company.—A provisional organization for this unit includes a headquarters platoon, a procurement platoon, and a warehouse platoon. Its mission is to establish and operate an engineer spare parts supply depot and other spare parts supply agencies. It may operate as an individual supply unit. Sections or detachments may assist in operation of supply points in army service areas and in corps.

d. Engineer maintenance company.—This company consists of a head-quarters platoon, two maintenance platoons, and a contact platoon. It executes third-echelon maintenance of all equipment for which the Corps of Engineers has maintenance responsibility. This includes engineer equipment used by other arms and services as well as that used by engineers. It is attached to corps and higher echelons.

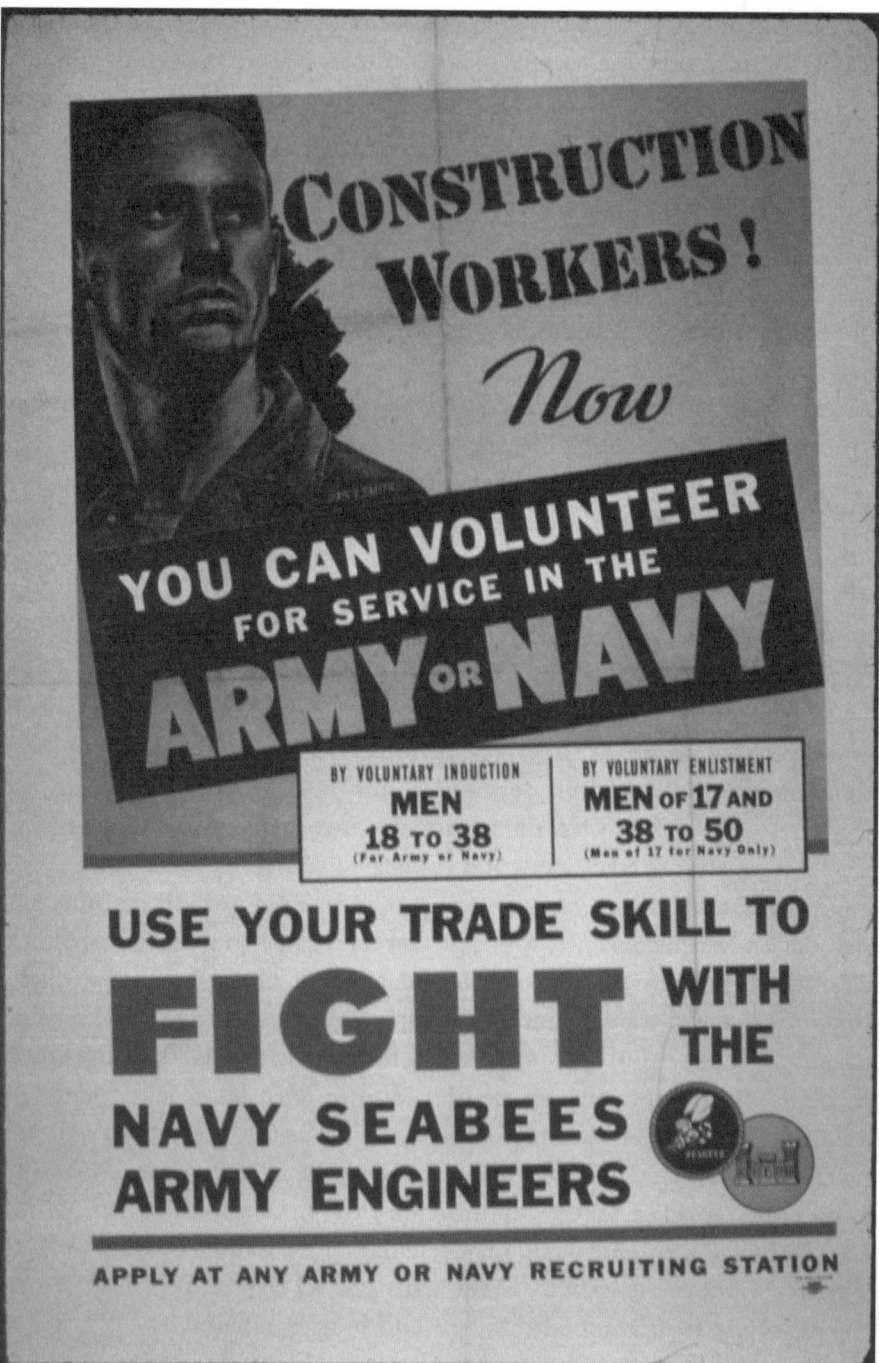

A recruitment poster shows how civilian construction workers can channel their skills into military engineering.

e. Engineer dump truck company (with Army Service Forces as well as Army Ground Forces).—This unit consists of company headquarters and two platoons of two operating sections each. It furnishes dump trucks to transport road and other construction materials on engineer work. Exclusive of dump trucks, the organic equipment is that necessary for maintenance of the unit in the field. It is attached to army and higher echelons.

f. Engineer topographic company (corps).—This company makes, procures, and reproduces maps for a corps. It increases the density of control and extends control for field artillery fire. It is equipped with mobile printing and photographic equipment mounted in van type trailers.

g. Engineer topographic battalion (army).—This battalion procures and reproduces maps for an army. It makes controlled and uncontrolled mosaics and maps of limited areas. It contains a survey company which may extend surveys in army areas to corps units. All equipment is mobile.

h. Engineer camouflage battalion (army).—This battalion may be attached to an army. It consists of a headquarters, headquarters and service company, four lettered companies, and a medical detachment. It gives technical assistance in camouflage methods and inspects camouflage work.

i. Engineer camouflage company (separate).—This company operates with an independent corps or task force. Its duties are similar to those of the army camouflage battalion.

j. Engineer water supply battalion.—This battalion consists of a headquarters, headquarters and service company, three lettered companies, and a medical detachment. It is equipped to procure, pump, purify, store, distribute, and transport water. It is attached to army and higher echelons.

■ 6. ENGINEER UNITS, COMBAT, WITH ARMY AIR FORCES.—*a. Engineer aviation battalion.*—This battalion consists of a headquarters, headquarters and service company, three lettered companies, and a medical detachment. It constructs independently an airdrome with all appurtenances. Organic equipment includes many items of heavy construction machinery not found in other engineer units.

b. Engineer aviation company.—This company is identical to the lettered company of the aviation battalion. It is used for small construction programs or for maintaining airdromes in remote localities. It may be reinforced with additional equipment and operators for particular missions.

Field engineering included many less-than-glamorous duties. Here U.S. Army engineers operate a mobile laundry. (PhotosNormandie)

e. Airborne engineer aviation battalion.—This battalion consists of a headquarters, headquarters and service company, three construction companies, and a medical detachment. Personnel, lightweight construction machinery, and supplies are transported by aircraft to the scene of operations. The unit is designed to provide quickly the minimum base facilities necessary for limited operation from an advanced landing field.

■ 7. ENGINEER UNITS, SERVICE, WITH ARMY AIR FORCES.—*a. Engineer aviation topographic company.*—This company consists of company headquarters, two drafting and mapping platoons, a reproduction platoon, and a geodetic control platoon. It prepares and reproduces maps, aerial photos, and aeronautical charts for the Army Air Forces.

b. Engineer air force headquarters company.—This company consists of an engineering platoon, a camouflage platoon and a reproduction platoon. It is attached to an air force headquarters or to an air service command, and executes technical missions in connection with activities of engineer aviation units. One or more such units is provided for each air force.

■ 8. ENGINEER UNITS WITH ARMY SERVICE FORCES.—All such units are classified as service troops. They may be organized, trained, and equipped to do either general or special engineer work.

a. The engineer general service regiment includes a headquarters, headquarters and service company, two battalions of three lettered companies each, and a medical detachment. The regiment is trained and equipped to undertake all types of general engineer work in the communications zone. It does not have sufficient motor transportation to carry all personnel.

b. The engineer special service regiment consists of a headquarters, headquarters and service company, two battalions of three lettered companies each, and a medical detachment. It is designed to undertake the more important engineer construction projects of a permanent nature. In strength it is similar to the engineer general service regiment. There is a high proportion of noncommissioned personnel who specialize in design, preparation of plans, and supervision of difficult construction work.

c. The engineer heavy equipment company consists of a headquarters platoon and a service platoon. Its principal mission is to make available to general engineer units various items of heavy engineer mechanical equipment and some skilled operators.

d. The engineer heavy shop company is composed of a headquarters platoon, a manufacturing platoon, and a repair platoon. Its mission is to do fourth-echelon maintenance of all equipment for which the Corps of Engineers has maintenance responsibility. The manufacturing platoon is equipped with heavy-duty fixed shops. Its repair platoon is equipped with mobile repair shops.

e. The engineer topographic battalion, GHQ, is about twice the strength of the army type topographic battalion. It reproduces maps on four 22- by 29-inch rotary presses with auxiliary equipment. The equipment is fixed.

The battalion also is equipped to prepare maps by photogrammetrical methods and it may, if necessary, advance horizontal and vertical control to the zones of army type topographic battalions. Frequently it may be required to reinforce these units.

f. The engineer port repair ship is divided into a headquarters section and an operating section. It maintains channel markings and other aids for pilots, and removes obstructions from channels or ship berths.

g. The engineer port construction and repair group is a special unit which includes a headquarters, headquarters company, engineer group, and the necessary additional operating personnel from available engineer and other service units. It does engineer work in the repair or rehabilitation of waterfront facilities and installations of ports of debarkation in a theater of operations.

h. The engineer petroleum distribution detachment is composed of a headquarters section and several operating sections. Its mission is to design, construct, operate, and maintain military pipe-line systems as a means for transporting, distributing, and storing gasoline in bulk in a theater of operations.

i. The engineer gas generating unit consists of a commissioned officer and a small group of occupational specialists organized as two similar sections. It produces and supplies oxygen, acetylene, and nitrogen gases.

j. The engineer utilities detachment is flexible in organization; number and composition of officer and enlisted personnel depend upon population, location, area, and facilities of the post or military installation the unit serves. [. . .]

k. The engineer forestry battalion includes a headquarters, headquarters and service company, three or more forestry companies, and a medical detachment. Its mission is to exploit woodlands in or near a theater of operations so as to provide an abundant and ready supply of forest products, especially lumber.

l. The engineer parts supply company is composed of a depot head-quarters staff section, a headquarters platoon, a procurement platoon, and a warehouse platoon. Its mission is to establish and operate an engineer parts supply depot and other spare parts supply agencies. The company may form part of an engineer supply depot or of the engineer section of a general depot, or it may operate as a separate supply unit.

m. The engineer parts supply separate platoon is composed of a depot headquarters staff section, a detachment headquarters, a technical section, and a warehouse section. Its mission and operation are similar to those of the engineer parts supply company.

n. The engineer mobile searchlight maintenance unit is composed of a small group of occupational specialists equipped with a motorized third-echelon electrical repair shop. It provides mobile third-echelon maintenance for searchlights.

■ 9. ENGINEER COMBAT GROUP HEADQUARTERS.—This is primarily a tactical command group, composed of a headquarters, headquarters company, and attached engineer units. Engineer combat group headquarters are organized, in general, on the basis of one per four engineer combat battalions or the equivalent in other engineer units. They are attached to corps and army, and through them the corps or army engineer exercises his control of engineer units.

■ 10. ENGINEER HEADQUARTERS. Engineer headquarters are provided for corps, army, communications zone and its sections when established, theater of operations headquarters, general headquarters, army air forces and separate air forces, task forces, base commands, and defense commands.

[. . .]

CHAPTER 2
SECTION V

ENGINEER REGIMENTAL, BATTALION, AND COMPANY ORGANIZATION

■ 30. REGIMENT.—*a. Organization.* Engineer regiments include both general and special service. They are organized into a headquarters and headquarters and service company, two battalions, a medical detachment, and attached chaplain.

b. Attachments.—An engineer regiment may have other engineer units or elements of other arms and services attached to it. The regimental commander coordinates the action of attached units with that of his own.

c. Regimental headquarters.—Regimental headquarters consists of the regimental commander and his staff. [. . .] In general, duties and responsibilities of regimental commanders in commanding and supervising operations are the same as those of the commander of a divisional engineer battalion except that the regimental commanders have no staff functions with higher command. Functions, duties, and relationships of staff officers of engineer regimental headquarters are the same, in principle, as those of the staff officers of battalion headquarters of divisional battalions.

d. Headquarters and service company.—Headquarters and service company of the regiment consists of company headquarters, which includes the command personnel for routine administration, messing, and supply of the company; a headquarters platoon, which furnishes enlisted personnel for the staff sections of regimental headquarters; and a service platoon, which furnishes transportation, special equipment and operators, and repair service for the entire regiment. Company headquarters also operates the regimental officers' mess.

e. Battalion.—(1) Battalions which are *components of regiments* consist of a small headquarters and headquarters detachment and three lettered companies. They are unsuited for independent missions away from their regiments unless provided with a provisional service unit. Such a unit would be made up of detachments from headquarters and service company.

(2) *Independent battalions and squadrons* are made up of headquarters, a headquarters and service company or troop, two or more lettered companies or troops, and a medical detachment. The battalion headquarters and service company or troop is organized into a company or troop headquarters and various functional sections without a platoon organization, and has functions similar to those of the headquarters and service company of a regiment.

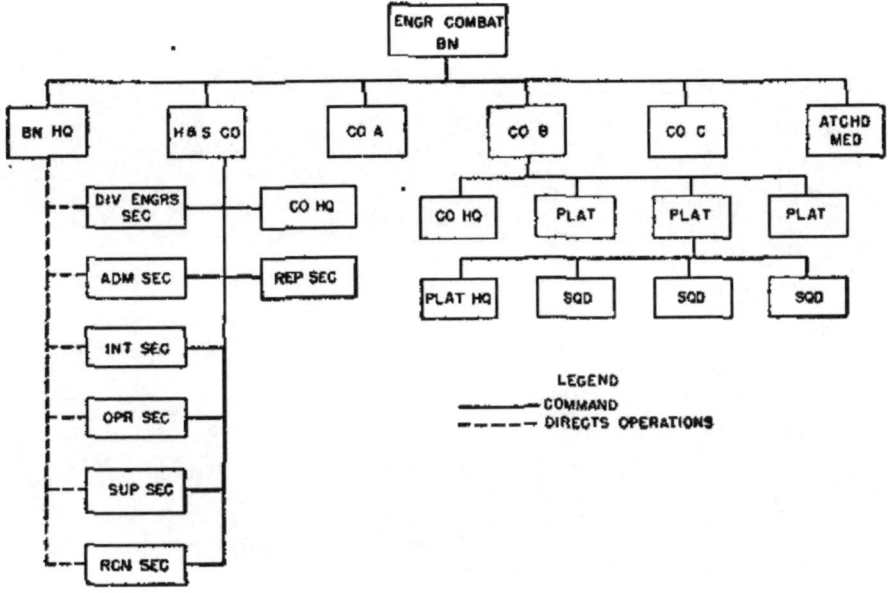

f. Attached medical.—(1) *General.*—Each regiment or independent battalion includes a medical detachment of two or more medical officers, one of which is dental. In the case of the mountain battalion one is a veterinary officer. The senior medical officer is the unit surgeon. Enlisted personnel include medical, dental, sanitary, and surgical technicians, truck drivers, litter bearers, and first-aid men. The mountain battalion also has veterinary personnel. For mess and routine supplies it is attached to one of the companies, usually headquarters company. Transportation of the usual medical detachment consists of a cross-country ambulance and light trucks and trailers for command, personnel, equipment, and supplies. Its routine equipment and medical supplies are sufficient for first-aid treatment and care of sick persons not requiring hospitalization.

(2) *Duties.*—The primary functions of the medical detachment are to collect casualties, and to set up and operate aid stations from which the more serious cases are evacuated by the division or higher unit ambulance services. The medical detachment operates most efficiently when used

Soldiers of the 41st Engineers on parade at Fort Bragg in North Carolina. (NARA)

as a single unit. It maintains a routine dispensary and sick call for minor ailments. [. . .]

(3) Since the medical detachment has neither the strength nor the equipment for adequate care of an engineer unit when dispersed, ordinarily engineers use the facilities of the nearest aid stations regardless of arm or service.

■ 31. COMPANY.—*a. General.*—The engineer company is designated as a lettered company or an independent company. In either case it is the basic administrative unit that can sustain itself in the field. It is the smallest engineer unit commonly employed on a detached mission.

b. Lettered company.—The lettered company is a component of an engineer battalion, and normally consists of company headquarters and either two or three numbered platoons. The following organization is typical of the combat company or troop:

(1) *Company headquarters* include the following personnel and their duties:

(a) A company commander, responsible for administration, discipline, training, messing, welfare, and operation of the company.

(b) An administrative officer, in most engineer companies a lieutenant in company headquarters who relieves company commander of normal routine duties of company administration, supply, messing, and operation and repair of transportation.

(c) A first sergeant, who assists company commander in administration of company, and is responsible for maintenance of all company records, interior administration, and police.

(d) A mess sergeant responsible to company commander for training of cook and cook's helpers, and charged with procurement and preparation of rations and feeding the men.

(e) A motor sergeant responsible to company commander for servicing, repair, and operation of all transportation assigned to company; and charged with training of all truck drivers, supervising loading and movement of company transportation, and directing activities of company automobile mechanics.

(f) A supply sergeant, who supervises activities of supply personnel and is charged with procurement and issue of clothing, armament, equipment, and supplies; maintenance of supply records; and storage and care of company stocks.

(g) Other noncommissioned officers, technical specialists charged with supervision of special operations and advising company commander on special problems.

(h) Technicians of third, fourth, and fifth grades, receiving pay allowances of noncommissioned officers of same grade. These men are skilled occupational specialists and include tractor drivers, automobile mechanics, and air-compressor operators.

(i) Privates, including privates, first class, or privates who have routine assignments as bugler, orderly, or cook's helper. Basics are used as fillers and replacements as circumstances require.

(2) The *platoon* is usually divided into headquarters and three squads or operating units. The headquarters usually includes a lieutenant as platoon commander, a platoon sergeant, and enlisted men for routine work. The platoon commander is responsible for training, discipline, and employment of his platoon. Actual handling of the platoon in engineer work is the duty of the platoon sergeant, the lieutenant remaining free to exercise general supervision.

(3) The *squad* is the basic operating unit and usually consists of a sergeant, unit foreman; a corporal, assistant unit foreman; and other enlisted personnel including carpenters, electricians, mechanics, riggers, demolition men, and truck drivers. In completely motorized units, the squad has organic transportation for tools, equipment, and personnel.

■ 32. INDEPENDENT COMPANY.—An independent company normally consists of a headquarters platoon and one or more other platoons. Generally they are organized, trained, and equipped to do special work.

a. Headquarters platoon includes company headquarters and one or more sections variously designated as administrative, service, supply, maintenance and repair or other overhead function for the benefit of the entire company. Headquarters platoon functions are similar to those of company headquarters of a lettered company.

b. Platoons of independent companies are normally divided into platoon headquarters and one or more operating sections, the headquarters supervising and controlling work of the sections. Each section has personnel and equipment for doing technical work or for executing special engineer missions.

c. Organization of independent companies is not uniform.

German Combat Engineers

An interesting comparison to the U.S. Army engineers during World War II is their German counterparts, the Pioniertruppen. The following extract comes from the U.S. War Department's Intelligence Bulletin of October 1942. Its key theme is a barely hidden admiration for the German engineer as a highly skilled combatant, rather than as a practical generalist. In part, this tighter skill set is delivered through intensive offensive and defensive training, but it also depends on the use of broader National Socialist labor forces to handle larger infrastructural engineering work:

1. INTRODUCTION

German combat engineers have been in the front lines of every major Nazi engagement of the present war. They form a very definite part of the German combat team, which also includes the armored forces, air forces, infantry, and artillery. The major duty of these engineers is to keep the German Army moving. They assault fortifications or other obstacles; they span streams with everything from log rafts to large temporary bridges; they go regularly into combat, and under the most difficult conditions, to clear the way for the echelons that follow. The success of the Germans through surprise, deception, and speed has been due in no small measure to the front-line work of the combat engineers, who during World War I worked in the rear areas except when called to the front under rare critical conditions.

2. TRAINING OF COMBAT ENGINEERS

In line with a training principle used throughout the German Army, 90 percent of the instruction now given to the army's combat engineers deals with attack problems and 10 percent with defense problems. Stress is placed on engineer reconnaissance and on making use of all means available in the combat area to help the German forces continue their advance. Army engineering equipment is used only when local means are not available.

The combat engineers are trained basically as infantry soldiers, since most of them now advance with the infantry and other combat troops and engage regularly in battle.

German military leaders, however, do not try to turn a combat engineer into a "Jack-of-all-trades." After receiving basic infantry and combat-engineer training, each trainee is assigned to a group which concentrates on one special type of work. For instance, he usually qualifies specifically for one of the following tasks: demolitions, fortifications, storm-troop combat, combat at rivers, construction of military bridges, emergency bridge construction, and general obstruction duties.

Some of these different kinds of specialized training are elaborated briefly as follows:

a. Storm-troop combat, with special equipment for rush assaults.

b. The obstruction service, which prepares obstacles of all kinds. These men are trained to handle explosives and mines, as well as to use electric saws and boring equipment.

c. Combat at rivers, which involves the use of rafts and small assault boats, both in the attack and in the defense. These men learn how to cross water under all conditions—in rain, heavy wind, and snow, and especially at night.

d. Construction of military bridges—also, establishing emergency ferry services, which provide transportation for men and materiel in motorboats and rowboats with outboard motors, or on improvised rafts propelled by these boats.

e. Emergency bridge construction, which calls for the preparation of many types of bridges, using material found locally.

f. Construction of field fortifications, which includes the building of defense installations of all kinds, large and small, with special training in the technique of preparing unusually deep foundations.

Looking over this whole set-up, the American soldier will see that a combat engineer in the German Army operates both as a fighter and a highly skilled technical expert. It may be said that if we destroy a German combat engineer, we destroy a man who is as useful to the Axis as any single person on the battlefield. Someone who is an infantryman—plus.

3. OTHER ENGINEER ORGANIZATIONS

The types of jobs done by combat engineers in World War I are now done almost exclusively by labor organizations, which include the so-called Todt Organization. The campaign in Poland taught the Germans that motor highways are likely to be more useful transport routes than railroads. The work of maintaining roads became so heavy that the government ordered a man named Todt, inspector of roads in Germany, to form a special organization for this important duty. The Todt Organization is composed of specialists and laborers who repair, construct, and maintain roads and bridges (with the Construction Engineers) from the rear and well into the combat zone. Sometimes it also assists in preparing fortifications (with the Fortification Engineers).

Another organization working with the engineers of the German Army is the Reich Labor Service. It trains boys of 17 and 18 to perform many of the tasks which in the last war were assigned to the regular engineers. These tasks include:

Constructing and maintaining important highways; constructing and improving fortifications, bridges, and airports; salvaging equipment, munitions, and materiel in battle areas, and camouflaging and sandbagging military establishments.

★★★

The following text once again comes from the pages of the Engineer Field Manual: Engineer Troops, *but with a change of focus to that of engineer troop training. The manpower demands of World War II presented a towering challenge to the training infrastructure of the USACE. The curricula for engineers were varied and highly technical, and delivery of those curricula had to be scaled up to meet the huge inrush of wartime recruits. By way of example, in July 1940 just under 10,000 men joined the engineers; in July 1941 that figure was 69,079 and that was before war had even begun for the United States. The Army was assisted in its recruitment efforts by the fact that some U.S. civil agencies had already given millions of young men a degree of labor and engineering experience in the 1930s and early 1940s. Chief amongst these was the Civilian Conservation Corps (CCC), which recruited men aged 17–28 to apply themselves to public works projects and conservation programs, giving them many skills the engineers would find useful, from forestry to electrical work, plus insight into working in "squads" and "platoons." A total of 3 million men served in the CCC, an invaluable human reservoir from which the engineers could draw. The USACE also ensured that relevant civilian occupations, such as electrician, plumber, carpenter, boilermaker, bricklayer, and steelworker, were identified and prioritized at recruitment stage, particularly once voluntary enlistment ceased in December 1942. The recruit would be sent onwards to one of several training centers stateside, where they would complete a 12–13-week (some schools later delivered 17-week courses) engineer variant of the Army Mobilization Training Program (MTP). As well as giving them the basic skills of soldiers, the MTP included core engineering training, with further advanced training available through specialist courses and in-service training.*

★★★

From FM 5-5, *Engineer Field Manual: Engineer Troops* (1943)

CHAPTER 4
TRAINING OF ENGINEERS

SECTION I
MOBILIZATION TRAINING PROGRAMS (MTP)

■ 41. GENERAL.—*a*. Mobilization training programs provide for basic training of the individual soldier. They are effective upon activation.

b. The 5-series of MTP is prescribed for engineer troops. The programs furnish a general guide for the balanced training of troops so they may be prepared to take the field on short notice.

■ 42. TRAINING OBJECTIVES.—The training of all engineer troops follows the basic instructions in FM 21-5 [see Chapter 2], with the objective of developing an offensive spirit in the soldier and the unit.

■ 43. BASIC TRAINING.—All engineer soldiers are given basic military training. (See FM 21-5.)

■ 44. ENGINEER TRAINING.—*a.* Military engineer training undertakes to qualify individuals for duties necessary to the engineer work for which units are responsible. Training varies with the type of organization. The procedure is first to train the individual in his assigned specialty and then to train groups of individuals in combined tasks under their appropriate commanders.

b. The objective of engineer training is to develop effective military operating units, such as squad, platoon, and company. Squads and platoons should be kept intact. These are elementary engineer teams, held together by the spirit of comradeship and the direct personal influence of their leaders.

c. Specialist training varies with the individual's prior experience. Technical specialists and machine operators in headquarters and service companies and in company headquarters are trained by attachment to companies and platoons doing work that demands exercise of their specialties.

■ 45. COMMAND TRAINING.—Although command training applies only to leaders, it is vital. Command should be decentralized and interference with subordinate commanders held to the minimum consistent with coordinated effort. Officers must know how to plan, estimate, organize, and supervise work, and how to allot tasks to subordinate units to insure that, without supervision, an assigned mission will be begun promptly and executed rapidly.

■ 46. COMBAT TRAINING.—*a*. Combat training instructs the unit in combat and in security when on the march, in bivouac, and while engaged in engineer work. Engineers use extended order drill as prescribed in FM 22-5. Infantry methods and formations prescribed in FM 7-10 are modified only to conform with engineer strength, armament, and organization. Any general procedures adopted by engineer units are uniform within all subordinate units.

b. Engineer troops are armed with a variety of weapons. Engineers must be trained to care for their individual and supporting weapons, to be proficient in their use, to know their capabilities, and to keep them clean and ready for immediate use at all times.

SECTION II
UNIT TRAINING

■ 47. GENERAL.—*a*. The primary purpose of unit training is to provide balanced training with emphasis upon well-organized and well-trained squads and platoons. Unit training programs are put into operation after completion of individual training.

b. This section is prepared as a guide to unit training of the engineer combat battalion. Its principles, modified by differences in organization, mission, and equipment, apply to other engineer units.

c. The unit commander supervises and inspects training to determine its progress and adequacy. Constant supervision is essential, but it must be informal and should not interrupt continuity of training. In addition, each commander, from the platoon up, must test his units to insure the mastery of each subject before passing to the next.

■ 48. SCOPE.—*a*. A unit training program is interested only in training units. Combined training follows. Enlisted men receiving unit training should have completed individual training and be reasonably well trained as individual soldiers and as basic members of the squad and platoon, ready to progress to more advanced training. Advanced training should include practical application of previous training. It should carry training of individuals, specialists, and small units to standards considerably higher than those reached at the end of individual training. It should prepare platoons, companies, and the battalion for combined training.

b. Divisional engineers need almost no unit training in general construction as practiced in the zone of the interior or in peacetime. Construction projects in camps, such as building barracks, target ranges, bayonet courses, and grenade courses, and normal peacetime road maintenance, while affording excellent training for other general engineer units, are not proper training assignments for combat engineer units.

c. During unit training, officers and enlisted men of the combat battalion should be developed into an engineer unit capable of performing any duty normally assigned to it in a combat situation.

d. Any training program, once decided on and started, may require modification for many reasons. However, progressive and balanced training should be preserved.

■ 49. TRAINING FACILITIES.—a. Site.—Problems of training combat engineer troops are similar to those common to other combatant arms and services [. . .] Training may be carried out in almost any locality, but if possible terrain and climate should approximate conditions in the probable theater of operations. Large areas are necessary; for the prescribed training subjects include not only most of those given infantry but also certain engineer subjects, such as explosives and demolitions, which require additional isolated areas as a safety precaution. The terrain should vary from flat to rolling or mountainous, contain numerous types of roads and bridges, have both dry and muddy ground conditions, and include sand, clay, loamy, and rocky soils. It should also contain all kinds and sizes of standing timber, and streams and gullies of various depths and widths.

b. Equipment.—Full use of all equipment must be made to insure efficient and complete execution of engineer training missions. When the amount of training equipment is inadequate for the number to be trained, schedules must be prepared for rotating the available equipment. When shortages in training equipment occur, substitute equipment should be obtained or improvised. Resourcefulness in devising training expedients is essential.

U.S. Army engineers operate a drilling rig during the construction of a U.S. field hospital to the east of Carentan, France, in 1944. (PhotosNormandie)

■ 50. TRAINING TEXTS.—*a.* For a list of War Department publications and visual training aids for conducting engineer training, see FM 21-6 and 21-7.

b. Normally, Field and Technical Manuals contain sufficient material for training purposes, but ingenuity must be used to apply this material to specific training tasks.

c. Training films, film strips, and graphic portfolios are valuable aids to instruction and are employed where practicable.

■ 51. TIME.—Unit training programs normally are based on a 48-hour training week. More time is utilized when desirable, especially in marches and field exercises. Open time is used to compensate for interruptions; for additional instruction in subjects inadequately learned; to provide refresher training; and for subjects given local emphasis, such as orientation talks. Formal athletic competitions, or preparation for athletic competitions should be conducted in time outside the prescribed training week.

■ 52. PROCEDURE AND METHODS OF INSTRUCTION.—Instruction is conducted as prescribed in FM 21-5. The subordinate unit commanders are the instructors in all unit training. Only in schools, review of MTP training, and other allied subjects, should instruction be centralized. Advanced training should be practical and should repeat MTP training only when necessary.

a. *Basic training.*—(1) Review of basic and general subjects must be continued regularly to maintain a high standard of individual proficiency.

(2) In order to avoid monotony, periods devoted to review of basic training should be short.

b. *Field exercises.*—Squads, platoons, companies, and battalions should cover an engineer subject by field exercises involving a tactical situation. Emphasis should be placed on solving the tactical as well as the engineer problems. Where applicable, training in the supply of engineer materials, rations, water, gasoline, and ammunition should form a part of the exercise. Periods should be long enough to permit the performance of the assigned task under the assumed tactical situation; frequently this will mean several days for a single exercise. Night operations should include technical operations, such as bridge building and laying mine fields, as well as tactical operations. One or more of the subjects included in the weekly program should be covered in a night operation. All operations should be conducted without lights. Every exercise should be followed by a thorough critique.

c. *Organizational unity.*—In combat zones, combat battalions normally accomplish their technical engineer tasks by breaking up into small units. Therefore the emphasis in technical training should be placed on making squads and platoons effective operating units. For bridging missions companies must be trained as units.

d. *Troop schools.*—Troop schools for officers and noncommissioned officers will be conducted throughout the entire training period as preparation for subsequent instruction of units. The schools should rehearse basic principles of each exercise before execution. The schools will be held as directed by higher authority.

e. Troops' preparation.—Troops should be grounded in the fundamentals of each task prior to its execution.

f. Performance.—Whenever any unit performs an exercise in an unsatisfactory manner, the exercise should be repeated until it is done satisfactorily.

■ 53. ENGINEER TRAINING.—The military application of all engineer work will be stressed constantly so all personnel may see it in its proper perspective.

■ 54. TACTICAL TRAINING.—The primary mission of the engineer combat battalion is engineer work. However, in an emergency the battalion may be held in mobile reserve and used as infantry in combat. Therefore tactical training must be conducted in order to meet that emergency.

a. Scope.—Tactical training that is stressed includes protection of working parties, defense of road blocks and other obstacles, combat actions of squads and platoons, motor movements, entraining and detraining, entrucking and detrucking, night tactical operations, field tactical operations, and field tactical training of the battalion.

b. Combat intelligence.—Each commander of a combat unit is responsible for obtaining information on the enemy forces opposing him. In general, in the combat battalion, the staff agency for combat intelligence is the intelligence (S-2) section. Personnel used for intelligence work are trained and employed in accordance with the doctrines prescribed in FM 30-5, 30-15, and 30-25. Since the reconnaissance section of the combat battalion is not available for combat intelligence most of the time a small group should be trained in this duty. Training of the engineer squad should include additional training for combat or reconnaissance patrol.

c. Security on the march.—The doctrine of security is found in FM 100-5 and 7-10 and should be followed at all times by all engineers. Moving columns make excellent targets for low-flying aircraft. All engineer troops must be instructed thoroughly in protective measures against such attacks.

d. Local security.—Engineer troops engaged in work at or near the front are trained to keep their weapons immediately available. During unit training, leaders will stress this practice and decide plans of action under simulated enemy interference.

■ 55. TRAINING IN CHEMICAL WARFARE.—*a.* To meet the probable use of toxic gas by the enemy, engineer troops must be trained thoroughly in chemical warfare; in how the enemy can use it; and in defense against chemical attack.

b. For thorough training, all instruction is practical rather than theoretical, making maximum use of the training munitions and supplies authorized by AR 775-10, Tables of Basic Allowances, and Tables of Equipment for engineers.

c. References.—Principles governing offensive and defensive use of chemicals, together with combined operations and security in connection therewith, are found in FM 100-5, 3-5, and TM 3-305.

■ 56. TRAINING IN OPERATION AND MAINTENANCE OF EQUIPMENT.—Motor maintenance, training of driver, and maintenance, care, and operation of engineer mechanical equipment are stressed concurrently with other training. Maintenance of organizational transportation and mechanical equipment is taught by training in—

a. First-echelon maintenance.—This training is essentially preventive maintenance by operator or driver and assistants. It includes—

(1) Correct operation.

(2) Operator servicing, lubricating, and cleaning.

(3) Tightening and minor adjustments.

(4) Inspections within companies and similar units.

b. Second-echelon maintenance.—This training is essentially preventive maintenance by organizational mechanics and maintenance specialists. It includes—

(1) Centralized organizational servicing and lubrication.

(2) Preventive maintenance, adjustments, minor repairs, and unit replacements within the limits of time available and equipment authorized.

(3) Systematic maintenance inspections within independent battalions and similar units.

■ 57. SPECIALIST TRAINING.—*a*. Technical or specialist schools will be conducted as necessary to perfect the individual in his technical specialty. Such schools should be scheduled to interfere as little as possible with unit exercises. Each individual specialist normally should attend any exercise scheduled by his unit.

b. Specialist training should include training of such communication personnel as radio and switchboard operators, message center clerks, and code clerks.

c. Technical engineer training, especially the military aspects of specialist work, is given maximum time. Because of the skilled nature of the tasks on which engineer units are employed, personnel should be recruited from men already trained in civil life for those tasks. A lack of equipment may make it impossible to give specialist training in early stages of training periods. But when equipment is available and units are operating under combat conditions, unit commanders should utilize the apprentice system continually to train additional personnel as replacements. Casualties cannot then cripple work of their units, and expansion of units to meet emergencies can be made readily.

■ 58. STAFF TRAINING.—*a*. Headquarters of the engineer combat battalion is perfected in its duties by training in—
  (1) Staff functions and operations.
  (2) Mechanics of issuing orders.
  (3) Planning of battalion operations.
  (4) Proper distribution of work to companies.
  (5) Inspection of the execution of engineer work.
  (6) Engineer needs of troops of other arms of the division.

b. (1) To plan and put into operation the unit training program of an engineer organization, battalion headquarters must develop and put into practice a standing operating procedure based on that of the next higher unit.

(2) The will of the commander should be expressed in standing operating procedure for technical and tactical emergencies. To be effective

it must be revised from time to time. Modern warfare is characterized by speed of movement and rapidly changing situations. No unit commander should permit a standing operating procedure to standardize the technical or tactical employment of his troops, to narrow the scope of training, or to destroy the opportunities for use of initiative.

■ 59. SERVICE TRAINING. Members of headquarters and service company are perfected in their specialties by additional technical instruction and training carried on concurrently with the performance of their duties. Training is scheduled for small groups when it will interfere least with the normal functional duties of the company.

■ 60. MEDICAL TRAINING.—Members of the medical detachment are perfected in their specialty by additional technical instruction and training conducted concurrently with the performance of their duties. In addition, training is given in basic, technical, combat, and tactical subjects. First-aid training is conducted under the supervision of the unit surgeon and is in accordance with the doctrines contained in FM 8-5 and other pertinent Medical Field Manuals.

Basic Skills and Engineer Equipment

FM 21-105, Basic Field Manual: Engineer Soldier's Handbook, *was the true back-pocket guide of every engineer in the U.S. Army, and hence is one to which we shall refer several times in this book. The purpose of the guide was clearly stated in its brief introduction, signed in June 1943 by order of George C. Marshall, the U.S. Chief of Staff: "FM 21-105, Engineer Soldier's Handbook, is published for the information and guidance of all concerned. Its purpose is to supplement FM 21-100, Soldier's Handbook, by giving the newly enrolled soldier of the Corps of Engineers, United States Army, a convenient and compact source of basic military engineer information and thus aid him to perform his duties more efficiently." Looking at the contents list of the book, it is revealing to see what fell under the category of "basic military engineering information," which included constructing field fortifications, applying camouflage, handing explosives, hunting and destroying enemy tanks, assaulting a fortified position, using boats and rafts, handling an engineering truck, and building an airdrome. If these were the core skills, then every engineer was to some degree a specialist within the U.S. Army.*

★★★

From FM 21–105, *Basic Field Manual: Engineer Soldier's Handbook* (1943)

CHAPTER 2
ENGINEER TOOLS AND COMMON ENGINEER TASKS

Section I
ENGINEER TOOLS

■ 4. IMPORTANCE.—*a.* The engineer soldier is an expert in many things. One of his most important skills is the use of many kinds of tools; some, hand tools, others, power tools. Tools are the basic implements of the engineer. They go along with his unit and are always at hand. With tools the engineer accomplishes many tasks. How well and how quickly he does his job depends upon—

(1) His skill.
(2) His physical condition.
(3) The condition of his tools.

b. All of these are the responsibility of the individual soldier. His own life and the lives of his fellow soldiers depend upon the tools and the skill with which they are used.

■ 5. CARE.—Mainly upon you, the soldier who uses these tools, depends the condition of the tools. When the supply sergeant or his assistant issues tools to you, you become responsible for them. Clean and oil them before you return them. If you are careful in the use of your tools, if you use them in the correct manner, if you are quick to notice and report such things as dullness, battered heads, and rough spots on handles, the job of keeping tools in good condition is easy.

■ 6. SAFETY.—Your tools are sharp. If they are handled improperly you or a comrade may be hurt. Learn to use your tools correctly; the correct way is both the easiest and the safest way. Here are a few general safety rules. Do not forget any of them.

a. Carry your tools properly.

b. Do not lay sharp tools, such as axes, adzes, and peavies, on the ground where they can be stepped on, fallen on, or run into.

c. When swinging a tool, make sure all others are a safe distance away.

d. Make sure all tool heads are tight on their handles.

e. Do not get in the way of another soldier who is using a tool.

■ 7. Use.—Tools are designed to do work with a minimum of effort. The untrained man tires himself by forcing his tools, gripping them too hard, or using an improper position. The trained man is relaxed, lets his tools do most of the work, and uses his mind, eyes, and hands to guide the tools.

Men of an Engineer Port Construction & Repair Group at Army Service Forces Training Center (ASFTC) Camp Gordon Johnston, Florida, build a floating dock from Navy pontoon gear, May 1944. (Signal Corps Archive)

■ 8. ENGINEER TOOL SETS.—Each engineer organization is equipped with the hand tools needed for accomplishing the work usually assigned to it. For convenience in selecting tools for a particular job, they are grouped into sets, such as carpenter, blacksmith, pioneer, and demolition sets. Learn to know the contents of the various squad and platoon sets.

■ 9. HAND TOOLS.—Most of the tools you use are hand tools, the most important of which are discussed below. These discussions are only a guide, however; they are not a substitute for actual training and extensive practice. Apply the things you read here at the first opportunity.

a. Ax.—Before starting to swing the ax, make sure that there is no interference in any direction. If there are overhanging limbs or undergrowth in the way, clear them out first. Make sure of a firm footing and see that no one is dangerously close. In swinging the ax, be especially careful to stand so that if the mark is missed, or if the ax glances off, it will not strike you. Keep your eyes on the point to be struck. Never throw the ax or leave it lying on the ground; instead, drive it into a log or stump, or put it in its box. Never use the ax to drive metal stakes.

b. Hatchet.—The hatchet is used for light trimming work such as framing timber, sharpening stakes, or splitting wood. The position of the hand depends upon the desired blow. Hold it near the end of the handle to strike a heavy blow for heavy cuts and near the head for light trimming strokes. The hatchet has a hammerhead which may be used for driving medium-size nails.

c. Adz.—The adz is a hewing and smoothing tool used by engineers mainly to remove bark and to square round timber. It must be used carefully or the user may be injured. The correct way to use the adz is to stand astride the log and take short hewing strokes. The log is first scored with chopping strokes, or with shallow cuts made with a saw.

d. Pick and pick mattock.—You should be able to use the pick or pick mattock with either the right or left hand leading. The pick is swung in a manner similar to that used in swinging the ax. To use it with the right hand leading, stand with your feet comfortably placed, left hand at the handle end, right hand near the pick head, body bent slightly forward, and arms hanging naturally. Carry the pick head behind and above your right shoulder without changing the position of your hands. Swing the pick head forward, allowing the handle to slide through your right hand until your hands meet, and continue the stroke downward. Keep your eye on the point to be struck.

① Start.   ② Top of swing.

③ Down swing.   ④ Swing completed.

Figure 3.—Using the ax on horizontal timber.

e. Shovels.—You should be able to use the shovel with either a right- or left-hand swing. After filling it by one of two methods, press the handle down and back to free the shovelful from the rest of the material. Then hold the handle down while raising the weight of the full shovel with the other hand. In casting, allow the handle to slide through the lower hand in the most convenient manner. Do not use a shovel as a pry.

f. Saws.—Saws are of various design, depending upon the kind of work required.

(1) *Hand saws.*—There are two kinds of hand saws—crosscut and rip. A crosscut saw has knifelike teeth and is used to cut wood *across* the grain. A ripsaw has chisellike teeth and is used to cut wood *with* the grain. The hand

saw is used in most common carpentry work. A saw cut should be started by guiding the blade against the thumb of the left hand and drawing the saw backward. Extending the forefinger along the handle aids in guiding the blade. Hold the saw lightly and do not try to push it into the wood; move it back and forth with a full, long stroke, letting it do its own cutting.

(2) *One-man saw.*—This saw is equipped with cutting and drag teeth and an extra handle so that, if desired, two men (one at each end) can use it. This saw is used on fairly heavy and rough timberwork where speed is more important than close fits or exact measurements.

(3) *Two-man crosscut saw.*—This saw has two removable handles and is used for cutting standing trees or for heavy framing or cutting. Two men operate it by pulling alternately. Do not push or "ride" the saw; one man's straight pull does the work while the other man relaxes but keeps his hand on the handle.

g. Clawhammer.—The clawhammer is used to drive and draw nails. In driving nails, the hand should be at the level of the nailhead at the moment of impact so that the nail is hit squarely and the force of the blow travels directly along the nail. Similarly, in drawing nails, the force should be directly along the nail.

h. Sledge.—The sledge is used for heavy driving, rock-breaking, striking rock drills, and for shop and general construction work. It should be swung like a pick. A full stroke gives best results.

i. Maul.—The maul is a heavy, wooden driving tool, and should be used only to drive wooden stakes and posts. It is swung like the sledge.

j. Peavy.—The peavy is a gripping and level-action tool, used to roll, haul, or carry heavy timber. To carry heavy logs with peavies, men should be distributed equally on each side of the log.

k. Bars.—There are several kinds of bars, of varied shapes: crowbar, wrecking bar, pinch bar. These are prying tools and are used as levers. In using these bars, secure as much leverage as possible and take small "bites" each time. Be satisfied with relatively small movement at the cost of little effort, instead of doing excessive work to make a large move.

l. Brush hook.—The brush hook is a sharp, curved cutting tool used to clear underbrush and to trim branches. It should be swung with both hands at the handle end.

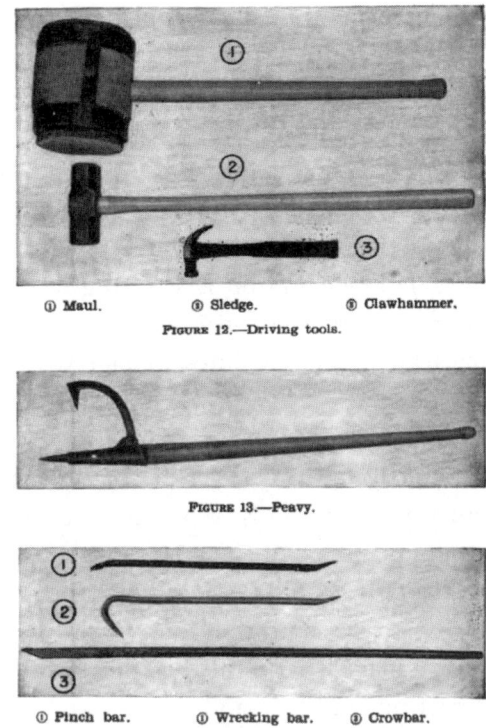

① Maul.   ② Sledge.   ③ Clawhammer.

FIGURE 12.—Driving tools.

FIGURE 13.—Peavy.

① Pinch bar.   ② Wrecking bar.   ③ Crowbar.

FIGURE 14.—Bars.

m. Machete.—The machete (pronounced muh-SHAY-tay, muh-SHET-ee, muh-SHET) has a long, extremely sharp blade with a wooden handle. It is used to clear underbrush and trim small branches. It is swung with one hand. Keep it in its sheath when not in use.

n. Earth auger.—The earth auger is an extremely useful hole-boring tool for the engineer. The 6- and 10-inch sizes are most commonly used. As far as possible, keep the cutting blades out of contact with rocks; use it with care in rocky ground.

o. Wire cutters.—Wire cutters are especially designed to cut barbed wire. The rubber-covered handles are insulated against live wires. The bent hooks on the searching nose are used to pull the wire toward the operator. The cutters are used with two hands.

p. Side-cutting pliers.—Side-cutting pliers are used both for holding and cutting, essentially with one hand.

q. Pocketknife.—The pocketknife has four blades, which include a combination reamer and leather punch blade, a screw driver and bottle-opener blade, a can-opener blade, and a cutting blade. It is equipped with a clevis for attachment to a carrying chain, thong, or cord.

r. Wrenches.—The two chief adjustable wrenches are the monkey wrench for angular bolts or nuts, and the pipe wrench for round fittings. Note the differences between them.

s. Brace and bit.—The brace and bit is a boring tool with a variety of bit sizes. It has different bits for wood-boring and for metal-boring. Make sure the wood drills do not come into contact with foreign material such as rocks and nails.

t. Ship-ring auger.—The ship-ring auger is a long boring tool used to bore holes deeper than those made with the bit and brace.

u. Plane.—There are several kinds of planes used for various kinds of work, but the working principle of all is the same. It is a smoothing tool with a fine cutting blade. It should be used with both hands guiding the stroke. Take long easy strokes with the grain of the wood, short strokes against the grain. Be especially careful that the blade is not nicked by nails or other obstructions in the wood.

v. Chisels.—(1) Wood chisels are struck with a wooden mallet, never with a metal hammer.

(2) Cold chisels for cutting metal are struck with metal sledges and hammers.

w. Measuring tapes.—The standard engineer measuring tape is a metallic, linen-fiber tape, rolled in a leather case. Keep it in its case when not in use. Since the metallic tape stretches slightly, the 6-ft. steel rule should be used for exact measurements.

x. Squares.—The try square is used to test square edges and surfaces. The steel framing square is used to measure angles and to draw the various lines needed by a carpenter.

y. Level.—The level is a precision instrument. By means of the bubble (bead) in the phials in the level, the engineer can determine whether or not a surface is horizontal or vertical. When the level rests on a surface and the bead is centered in the tube, the surface is level.

■ 10. POWER-DRIVEN TOOLS.—The power-driven tools used most frequently by the engineer are tools driven by compressed air from the mobile air compressor unit. These tools save much time and labor, and each engineer soldier should know how to use them. The tools most commonly used are clay diggers, wood and rock drills, pavement breakers, hammers, and wood saws.

SECTION **II**
COMMON ENGINEER TASKS

■ 11. MATERIALS.—Certain prepared building materials, such as standard-size lumber, are available to engineers at supply depots. However, very often the engineers must build their bridges, emplacements, etc., out of local materials found at the site of the work. Therefore, an engineer soldier must be always alert to note local materials, resourceful in his use of these materials, and quick to use them whenever he can.

■ 12. FELLING TREES.—With an ax, cut a deep notch near the base of the tree on the side toward which the tree is to fall. Then saw the tree on the opposite side to cut the remaining fibers, using steel wedges, if necessary, to keep the saw from binding. To cut the trunk clear of the stump, the saw cut should be started opposite the point of the notch. Where it is desired to keep the base of the tree firmly attached to the stump after felling, as in making a tree road block, the saw cut should be made considerably higher than the notch, so that all fibers will not be severed when the trunk falls. It is often advisable to use guy lines to guide a tree in falling and sometimes to use hand or motor power to pull a tree in a desired direction.

■ 13. MAKING TIMBER JOINTS.—In rough carpentry work, the butt joint and the lap joint are used to join or splice two pieces of wood so that they form one continuous piece. The butt joint requires the use of fishplates to hold the ends together. The lap joint is made by overlapping the ends of two timbers and nailing them together. This is the simplest and quickest splice for bracing and like uses.

■ 14. Driving Driftpins.—Driftpins (heavy iron spikes) are used to fasten large timbers together. Since driftpins are made of relatively soft iron, holes must first be bored in the wood before the pins are driven. These holes should be slightly smaller in diameter than the pin itself; for example, the hole for a ½-inch driftpin should be made with a ⁷⁄₁₆-inch bit.

■ 15. Handling Loads.—*a. Heavy lifts.*—The proper method of lifting heavy loads is to make the legs do the work. Do not bend over from the waist and throw all the strain on the groin and back muscles. Improper methods of lifting often cause a hernia (rupture).

b. Carrying long or heavy loads.—(1) For long, fairly light objects, such as timber beams or ponton balk, one man takes each end; to keep it from tipping over, the load rests on the right shoulder of the man in front and on the left shoulder of the man in the rear.

(2) For carrying somewhat heavier objects, more men may be used in a similar manner, but it is better to use pick handles, pipes or bars of ample length placed underneath. Two men (on opposite sides of the load) carry each handle. Timber and rail tongs, if available, should be used in the same manner, except that the load hangs below the handles. Extremely heavy loads should be handled on pipe rollers, wheeled dollies, block and tackle, or by machines.

(3) For small but heavy loads, a wheelbarrow should be used with the load placed evenly as far forward as practicable.

c. Carrying chess.—A wide one-man load, such as plank or 10-ton ponton chess, is carried on edge, rear end down, next to the body, with the right hand underneath, near the middle or balance, and the left hand on top steadying and guiding the load. Sometimes, when the chess is unusually muddy and slippery, or when fatigue necessitates the use of two supporting hands, the chess may be carried with both hands underneath. When this is done, however, special care must be taken to control the plank so that no one is hit by the ends. For carrying 25-ton ponton chess, two men are needed.

■ 16. Using Sandbags.—Sandbags are always laid with the chokes (mouths) tucked under and the side seams and tied ends inside. Grain, cement, and similar bags can be used, but they should not be more than half-filled or they will be too heavy for a man to handle. Sandbags are used frequently as reveting material to bolster the sides of holes in the ground. [. . .] To lay sandbags properly, they must be shaped so that when in place they are roughly half as wide as they are long.

★★★

U.S. Army engineers in World War II were kitted out with the same weaponry, personal equipment, and uniform as the broader U.S. infantry. What set them apart was the selection of essential maintenance and labor tool sets issued at squad and platoon levels, each set coming packaged in plywood cases. So, for

This 1,600ft mountain tramway was built in mountainous Italy by engineers of Company D, 126th Mountain Infantry Engineers, 10th Division, in just nine hours. (Signal Corps Archive)

Two Army engineers from the 829th Engineers work in a mobile machine shop in England in 1943. (U.S. Army Signal Corps, Pearson)

example, the "Engineer Pioneer Equipment Squad Set No. 1" consisted of axes, picks, shovels, machetes, brush hook, post hole auger (a screw-type device for removing plugs of earth when erecting fencing posts), sledgehammer, jack, and other selected small tools. Other sets would contain tools such as crosscut saws, crowbars, chisels, levels, pole climbers, and bench grinders, while powered tools at platoon, company, or battalion levels included the Model 6 chainsaw, pneumatic drill and tamper, block and tackle, and A-frames. The following text, from FM 5-5, Engineer Field Manual: Engineer Troops *(1943) summarizes the principles on which both minor and major equipment was issued. Note the reference to "chemical warfare equipment." The responsibility for chemical warfare originally belonged to the 30th Engineer Regiment (Gas and Flame) in World War I, but was eventually given to a new formation, the Chemical Warfare Service, in June 1918, a service headed by an experienced engineer officer.*

★★★

From FM 5-5, *Engineer Field Manual: Engineer Troops* (1943)

CHAPTER 3
EQUIPMENT AND ARMAMENT OF ENGINEERS

■ 33. GENERAL.—*a. Basis.*—Equipment issued to engineer organizations is prescribed in tables of basic allowances (T/BA-5) or tables of equipment (T/E-5 series). These are supplementary to AR 310-60, in which general provisions governing their preparation and application are given. For each supply service there is a section in T/BA-5 or the T/E-5 series which lists items furnished by that supply service, and basis of issue per organization or subdivision thereof, or per individual. Supply services which prescribe allowances of equipment for engineer troops are the Chemical Warfare Service, the Corps of Engineers, the Medical Department, the Ordnance Department, the Quartermaster Corps, and the Signal Corps. Modifications in the issue of equipment are made from time to time in accordance with developments in weapons, equipment, and organization; critical shipping requirements; and changing conditions of warfare.

b. Arm and service publications.—Components of sets and kits, spare parts and accessories to articles, and supplies issued to organizations on a time basis are found in supply publications of the supply services. For engineer supplies the publication is Engineer Supply Catalog, Parts I, II, and III.

c. Organizational balance.—In the preparation of engineer tables of organization and engineer tables of equipment a balance has been maintained among the controlling factors of personnel, duties, equipment, weights, and mobility. Weights of tools, accessories, spare parts, and organizational equipment have been kept within the safe capacity of the assigned trucks and trailers. Special heavy equipment and machinery are organically assigned to engineer units or carried in depots in a theater of operations for issue when necessary. This issue is affected by the nature of the engineer work to be accomplished, the importance of time for completion of the work, and the capacity of the engineer unit in trained personnel and repair facilities fully to utilize the maximum performance of the equipment. In order to maintain organizational balance, the factors outlined above must be considered along with the special requirements for each theater of operations.

■ 34. ORGANIZATIONAL EQUIPMENT.—*a. General.*—Organizational equipment consists of individual clothing and equipment, messing equipment, marking and cleaning kits, and other standard sets issued generally to similar units of all arms. Since issue and purpose of such equipment is the same for all engineer units, they are not discussed in subsequent chapters but are summarized below.

b. Companies.—Companies are issued organizational equipment for interior administration; for messing, sheltering, supplying, and otherwise providing for the men; and for the maintenance and repair of transportation and equipment. Functional equipment is discussed in subsequent paragraphs.

c. Regiments and independent battalions.—Additional organizational equipment is not issued to the headquarters itself, but to pertinent headquarters and service companies or troops. It includes additional tentage, officers' mess equipment, field safes, typewriters, duplicating machines, and similar administrative accessories.

d. Engineer headquarters.—Engineer headquarters is issued organizational equipment necessary for enlarged administrative duties; for shelter, messing, and care of the unit engineer and his staff; and for enlisted personnel of the headquarters. The basis of issue in the tables of basic allowances is the number of officers and men comprising the particular headquarters. The items of equipment correspond generally to those issued to companies and headquarters and service companies for similar purposes.

e. Use.—Organizational equipment as a rule is used by special personnel included in company headquarters. Thus the first sergeant is assigned and held responsible for administrative equipment such as field desks, typewriters, and other equipment used by company clerks, stenographers, and orderlies. Similarly, the mess sergeant is responsible for items such as field ranges, cooking utensils, and kitchen tentage; and the supply sergeant, for stock of spare parts and equipment, for marking and cleaning equipment, and for repair kits.

■ 35. ENGINEER EQUIPMENT.—*a. General.*—Tools, machinery, and transportation are equipment primarily for engineer work and differ with each engineer organization. As a general rule tools and machinery used by engineer troops are standard commercial items.

b. Basic sets.—For convenience of issue, engineer tools have been assembled into sets suitable for different classes of work. Basic tool sets include blacksmith, carpenter, demolition, pioneer, and tinsmith. These sets are made up of essential hand tools and a small amount of miscellaneous materials required for the usual types of engineer work. Other basic sets are issued for drafting, duplicating, sign painting, and sketching work. All basic sets are issued to general engineer units, and some are issued to special engineer units.

c. Supplementary sets.—In addition to the basic sets, supplementary sets are issued to specific engineer units. These sets augment the equipment in basic sets, so as to give each engineer unit tools to suit its strength, and to enable it to do a wider variety of engineer work.

d. Mechanical equipment.—Mechanical equipment consists of power and construction machinery. All general and some special engineer units are authorized various items of this equipment. It enables small numbers of engineers to execute greater missions, and to accomplish missions in less time. Engineer units are organically equipped with labor-saving machinery and mechanical devices to as great an extent as practicable, consistent with mobility. Certain items of equipment are issued to lettered companies; other items are issued to headquarters and service company, either for performance of highly technical work that forms a small but important part of the engineer mission, or for assignment to subordinate units for reinforcing purposes. Mechanical equipment may be classified as—

(1) *Standard,* including pieces of mechanical equipment such as medium tractor with angledozer, motorized air compressor, road grader, concrete mixer, motorized earth auger, and powered shovel.

(2) *Special,* including pieces that are organic equipment with engineer aviation units and that are designed to expedite construction, maintenance, and repair of airdromes. Items of special mechanical equipment are carryall scraper, trencher, sheepsfoot roller and asphalt mixer.

e. Special equipment.—Special engineer units are issued equipment designed specifically for the task for which they are organized. Thus, forestry units are equipped with portable sawmills, ponton units with floating bridge equipage, mapping units with surveying and map reproduction equipment, and water supply units with well-drilling machinery and mobile water purification trucks.

■ 36. TRANSPORTATION.—*a.* Transportation organic with all engineer troop units and engineer headquarters is motorized, except in pack companies of the engineer mountain battalion; it has riding horses and pack and riding mules.

b. Vehicles issued to engineers are in accordance with current tables of basic allowances. They include engineer special-purpose trucks and trailers, ordnance trucks, trailers, halftrack vehicles, and scout cars.

Engineers of the 1056th Engineer Port Construction & Repair Group operate a D6 bulldozer during work in Cherbourg in 1944. (PhotosNormandie)

■ 37. ARMAMENT.—*a.* Engineer troops are armed for—

(1) Personal protection.

(2) Security.

(3) Tank hunting.

(4) Combat as infantry in an emergency.

(5) Protection and defense of barriers, mine fields, prepared demolitions, and obstacles.

(6) Defense of airdromes, dumps, depots, and other installations.

b. Individual weapons include bayonet, pistol, carbine, rifle, caliber .45 submachine gun, antitank rifle grenade, and hand grenades.

c. Supporting weapons are crew-served; they include caliber .30 machine guns, caliber .50 machine guns, and antitank rocket launcher.

d. The portable flamethrower is an assault weapon used against fortifications.

■ 38. SIGNAL COMMUNICATION.—*a. General.*—The term "signal communication" includes all means and methods used to send messages. The normal agencies of signal communication available to engineer units in their operations include the message center, a field telephone system, radio equipment, pyrotechnics, and a messenger service. Messenger service employs available light transportation, such as ¼-ton trucks and ¾-ton command cars, and runners; in the pack company of the mountain battalion animals are used. Signal equipment issued to engineer units is discussed below.

b. Radio. Radio communication equipment is issued to all Army Ground Forces engineer combat units and to engineer aviation battalions. It is organic with headquarters and service companies of the battalions of all of the above units, and with lettered companies of the armored engineer battalion only. In divisional units, one set is for operation in the divisional command net, the others are distributed to other echelons as directed by the commanding officer. Although the portability of radio sets favors their use by engineers, the sets should supplant other types of communication only in emergencies.

c. Wire communication.—(1) Signal equipment used in establishing wire communication includes switchboards, telephones, and accessories.

This equipment is issued to the headquarters and service company of engineer units.

(2) Engineer regiments establish wire communication from the regimental command post to the battalion command posts or to an advanced message center located as near as possible to the command posts of the battalion. An engineer unit with a brigade or division has wire communication established to the unit command post by brigade or division signal personnel.

Wire communications from battalions to lower units are established by personnel in the engineer unit. The heavy ponton battalion and light ponton company establish wire communications when constructing floating bridges and for traffic control on bridges.

d. Other signal facilities.—Other signal facilities include panels, flares, Very pistols, and signal lamps for air-ground recognition.

A Quonset hut, originally used as a barracks for the 736th Engineers, is relocated by engineers on occupation duties in Japan in the immediate years after the war. (U.S. Army Corps of Engineers)

■ 39. Air Forces Equipment.—Airborne engineer units are provided with cargo parachutes, individual parachutes, and aerial-delivery type container assemblies.

■ 40. Chemical Warfare Equipment.—For chemical warfare the principal items issued to engineer troops, in addition to portable flamethrowers, are service gas masks, decontamination apparatus, incendiary grenades, and chemical land-mines. Special gas masks are issued for animals of the pack companies of the mountain battalion.

Japanese engineers

From 1942, the U.S. forces in the Pacific came face-to-face with the work of Japanese combat engineers. The Japanese engineers demonstrated particular skill in the domain of amphibious operations, supporting landings on hostile coastlines or implementing river crossings. In the following extract from the October 1944 edition of Handbook on Japanese Military Forces, *the writers outline some of the equipment used in these operations:*

Chapter X: Equipment

Section V: Engineer Equipment
1. GENERAL. a. Japanese engineers are well-equipped and are armed as infantry. They have shown outstanding ability in both the construction and demolition of bridges. On the other hand, airfields and roads so far encountered have not been up to Allied standards in speed of construction or serviceability. This may be attributable to the fact that the Japanese have depended more on manual labor than on heavy equipment, which they have not taken into forward areas in any quantity.
b. The construction of field fortifications has been very highly developed, and even at remote points Japanese engineers have been successful in constructing first class defense positions from material immediately available.
c. Engineers are also well-equipped with a wide variety of explosive charges and other material for assault and demolition tasks.
d. The shipping engineers (*Sempaku Kohei*) are specially trained and equipped to operate a large variety of transport craft, including landing barges.
[...]

2. AMPHIBIOUS EQUIPMENT. a. Bridges. (1) *Assault bridges.* Several different models have been developed and standardized. Two types are illustrated in figures 379 and 380. One type is made of lengths of steel tubing, supported by bags filled with kapok. The sections are joined together and afterwards locked. They are light enough to be carried easily by foot soldiers. Crossings of streams 100 feet and more in width are reported possible with this type of bridge.

(2) *Ponton bridges.* The heavier bridge is suitable for artillery and heavy equipment. The boats which support it are of standard sizes, especially developed for this work. One type, designed for transport by wagon, has 2 bow sections, each 8.7 feet long, and 2 center sections, each 7.1 feet long. This boat weighs 1,650 pounds complete. An even larger boat of this type, which also comes in 4 sections, is 45 feet long and weighs 6,800 pounds complete. Another type is designed for packing by horses; it has 2 bow sections, each 4.4 feet long, and 3 center sections, each 4 feet long. The complete boat weighs 921 pounds, and is slightly over 20 feet in length. An even lighter version also exists.

(3) *Improvised trestle bridges.* The Japanese are skilled in the construction of wooden trestle bridges (fig. 381) which they erect with great rapidity from materials prepared beforehand or available locally. Joints usually are lashed with straw rope, and occasionally are strengthened with iron pins. Such trestles are found serving as approaches to ponton bridges in wide river beds; in shallow rivers they may be several hundred feet in length. Despite their flimsy appearance they are capable of supporting artillery and other heavy equipment.

(4) *Sectionalized steel bridges.* Prefabricated steel bridges are used by the Japanese, but not as widely as by some other Armies. One truss–construction, portable, steel bridge is 48 feet long and weighs 820 pounds.

b. Assault boats. (1) *Collapsible boats.* Several types of collapsible boats have been developed. One of these, model F (fig. 382) is an outstanding example of assault boat design, and is very widely utilized. The boat, divided into two sections, each of which collapses flat on itself, (fig. 383) is individually floatable. Each section is 13.6 feet long, 4.75 feet wide, and 2.18 feet high. The wooden frame is braced, and all joints are bonded with rubber. The boat will hold 20 men, and it is estimated that 9 such boats could be loaded flat on a 2 ton truck. Light outboard motors have been used to propel this boat. Three types of rubber (pneumatic) boats, of from 1- to 10-man capacity, are in use. These are similar in construction to the rubber boats used by other Armies.

(2) *Demountable boats.* Japanese engineers also operate a variety of demountable motor boats, fitted with outboard and inboard motors of various kinds. One small, 30-foot boat breaks into 4 sections and is propelled by an outboard motor.

Some of the outboard motors are arranged for animal pack. Another larger type breaks into only 2 sections; the stern section is fitted with a 4-cylinder, inboard gasoline engine of 30 horsepower. It is not believed that any of these demountable boats permanently mount any weapons.

c. Landing barges. (1) Japanese engineers operate a large variety of landing barges, which have been employed extensively in various theaters. Since Allied air superiority has seriously interfered with Japanese use of transports in many theaters, landing barges generally have been employed for the supply and evacuation of their forward areas. As many as 500 of these craft have been found congregated in one port.

(2) Some of the design features of these small vessels are of interest. The landing-barge screw shown in figure 384 is designated for operation in shallow waters and affords maximum protection to the screw. The Japanese generally are credited with the development of the folding ramp, which now is used so extensively.

(3) A typical landing barge is the Daihatsu Model A (Army).

★★★

Much of the work performed by the USACE during World War II was as unglamorous as it was critical. Over its 446 pages, FM 5-10, Engineer Field Manual: Communications, Construction, and Utilities *(1940) lists and explains many of the engineers' core construction and maintenance activities, many of them as applicable to civil works as to wartime duties. Even the short section below gives a clear suggestion of the importance of engineer work in supporting the core functionality of a field army, from its sanitation and housing to its very ability to move across a theatre of operations. The constraints of space here mean that much has been omitted. For example, one specialist engineer duty was the "the construction, maintenance, and operation of military railways in the theater of operations" (p. 214). This service not only involved the adaptation of existing rail lines to military use or the laying of new tracks, but also the establishment of railway yards, rail storage depots, facilities for handling rail-moved personnel and animals (e.g., barracks, stables), signaling systems, and engine repair facilities. Naturally, this subject, as with many others within the engineers' remit, required a high degree of technical expertise, and those with prior rail engineering experience were welcome recruits.*

★★★

From FM 5-10, *Engineer Field Manual: Communications, Construction, and Utilities* (1940)

CHAPTER 6
CONSTRUCTION IN WAR

■ 176. GENERAL.—*a. Simplicity.*—All construction in the theater of operations is of the simplest nature. Only common construction materials and supplies should be used.

b. Economy of materials.—Economy of materials is obtained by—

(1) Limiting all construction to temporary, emergency facilities, providing only the barest necessities.

(2) Maximum use of existing structures and local materials.

(3) Use of type plans.

c. Type plans.—Type plans, such as those illustrated in this manual permit efficient utilization of personnel, materials, and available time. Detailed type plans are prepared in peacetime by the Corps of Engineers. They are the basis for the procurement of standardized materials to be shipped to the theater of operations.

d. Provision for expansion.—Projects should be laid out so that future expansion is feasible if any possible need for expansion can be foreseen. So far as practicable the project should be planned in its entirety and a suitable site chosen. Only units actually needed are constructed initially. They are practically completed before beginning additional work, even though an uneconomical working schedule results. The object is to obtain complete units for early use and to avoid work on units which later changes in plan may cause to be abandoned.

[. . .]

■ 177. SCOPE AND SITES.—*a. Scope of engineer construction.*—Among the facilities to be constructed by the engineers in the theater of operations are the following which are generally covered in this volume:

(1) *For use by two or more arms or services.*

(*a*) Semipermanent camps at ports, training centers, and rest areas.

(*b*) Ports.

(*c*) General depots.

(*d*) Administrative facilities.

(2) *For the Medical Department.*

(*a*) Station hospitals.

(*b*) General hospitals.

(*c*) Veterinary hospitals.

(3) *For the Quartermaster Corps.*

(*a*) Quartermaster supply depots, including gasoline and oil-storage facilities.

(*b*) Commissaries, including warehouses, bakeries, coffee-roasting plants, and cold-storage plants.

(*c*) Remount depots.

(*d*) Motor transport depots.

(*e*) Delousing plants.

(*f*) Laundries and dry-cleaning plants.

(4) *For the Ordnance Department.*

(*a*) Ammunition depots.

(*b*) Repair and maintenance depots.

(*c*) Ordnance supply depots.

(5) *For the Air Corps.*

(*a*) Airdromes.

(*b*) Repair depots.

(*c*) Assembly plants.

(*d*) Air Corps supply depots.

(*e*) Training centers.

(6) *For the Corps of Engineers.*

(*a*) Engineer supply depots.

(*b*) Railway facilities.

(c) Utilities.

(7) *For other arms or services.*

(*a*) Arm or service supply depots.

(*b*) Miscellaneous small installations.

b. Reconnaissance.—Features to be considered in reconnoitering for construction sites are tabulated below.

TABLE XLV.—*Construction sites*

FEATURES APPLYING TO ALL TYPES
OF CONSTRUCTION SITES

(1) Sufficient size for present needs and future possible expansion.

(2) Adequate water supply.

(3) On or near railroad of sufficient capacity for supply and personnel movement.

(4) Available for lease (if not already owned or leased by the Government for period up to 5 years.

(5) Largely free from floods.

(6) Adequate drainage.

(7) Roads good or potentially good.

(8) Climate favorable.

(9) No insect pests.

(10) Location strategically convenient.

(11) Material and labor locally available at reasonable prices.

ADDITIONAL FEATURES APPLYING
TO CAMP SITES PRIMARILY

(12) Accessible to adequate training area.

(13) Accessible to suitable target-range area.

(14) Recreational facilities nearby.

(15) Grazing facilities for animals. (Applies also to remount-depot sites.)

■ 178. ORGANIZATION.—Construction of a large project requires an adequate overhead organization. Assignment of working personnel to the job is governed usually by the receipt of materials. The working force should be initially small and then should be increased as construction progress demands. Construction in the theater of operations may be either with troop or civilian labor. The latter may be either hired labor under military supervision or labor employed by a private contractor. The organization chart in figure 95 is generally suitable for large construction projects prosecuted with troop labor.

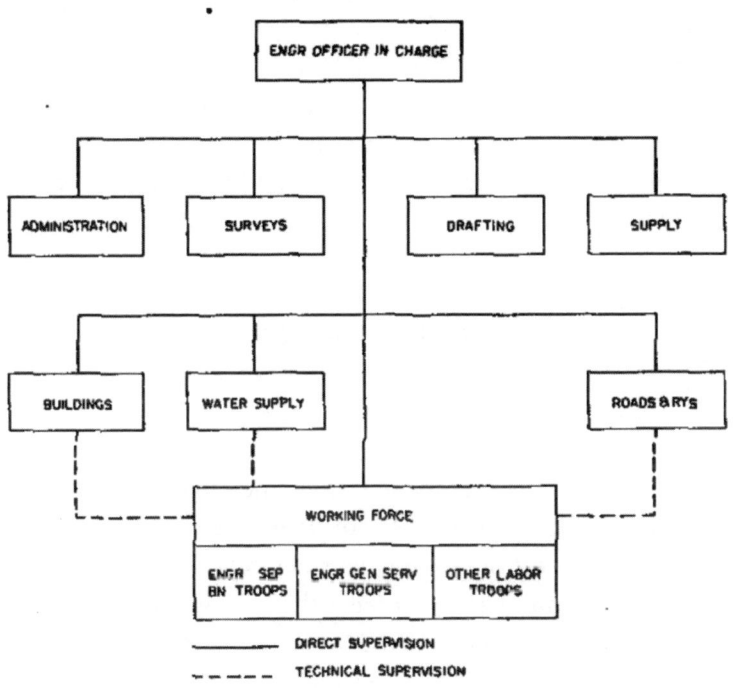

FIGURE 95.—A typical organization chart for a large project.

■ 179. SUPPLY.—*a. Local material.*—Local material sufficient for construction of a large project rarely will be available. Local sources should, however, be exploited to capacity, by such means as milling of local timber and development of quarries.

b. Shipments.—The majority of supplies must be procured and transported to the construction site. Great difficulty is frequently caused by temporary lack of nails, bolts, and similar items. In shipment, rails should be accompanied by bolts and spikes, roofing by nails and tacks. Structures should be shipped complete, because field classification and assembly are difficult.

■ 180. Air Defense Measures.—Defense of military facilities from aerial attack is easier if certain provisions are made in construction. The usual means of defense are here briefly discussed.

a. Dispersion.—Intervals between buildings or supply piles, which are standard precautions against fire, limit the amount of damage from aerial bombing.

b. Concealment and camouflage.—Installations often may be partially hidden or disguised, camouflage measures and natural concealment being used. Dummy installations aid in deception.

c. Active defenses.—Active defense is provided by antiaircraft weapons and by friendly aviation.

Section II
TROOP FACILITIES

■ 181. Standard Building.—*a. Discussion.*—Figure 96 shows the standard 20- by 100-foot building. This building is designed for a variety of uses, such as barracks, warehouses, mess halls, administration buildings, infirmaries, hospital wards, etc. It is essentially a lightweight frame, sheathed with wood or corrugated steel. Standard sizes of lumber are used. Bracing is limited to essentials: the stability of the structure depends partly upon the stiffness of the complete assembly of sides and ends. Corrugated steel is the simplest covering, but should not be used in hospitals because it is hot in summer and cold in winter. Common batten doors are used, with any simple available hardware. They may be single or double, covered with either corrugated steel or wood. Screen doors are used only on hospital wards, kitchens, and mess halls. The window frame, assembled in the field, is either screened or covered with a translucent material. Glass is not used. Ventilators, when required, may be either the ridge type or the tubular metal type. Floors are used only when absolutely necessary. The type A floor is installed on level and the type B floor on uneven ground. Use of the latter should be avoided, as it is more costly.

[. . .]

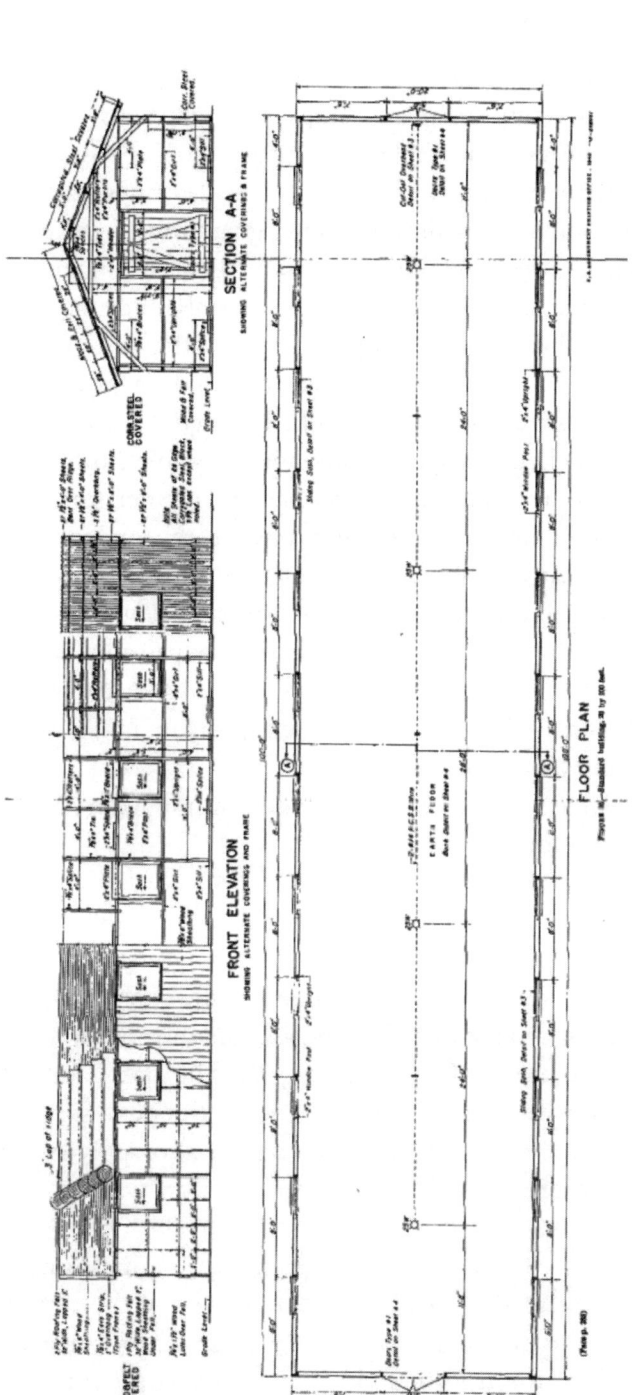

■ 182. Layout Of Semipermanent Camps.—*a. Requirements.*—Structural requirements for semipermanent camps for units for various arms will depend upon the current Tables of Organization. Facilities needed in every semipermanent camp are barracks, messes, latrines, baths, lavatories, administration building, medical building, guardhouses, storehouses, post exchange, officers' mess, and officers' quarters. Units having motors require shops; units having animals require stables, corrals, and watering troughs. One recreation building per regiment or independent battalion is conducive to good morale. Tents may be used in place of barracks. Local facilities, especially roads, should be used to avoid unnecessary construction. Kitchens, hospitals, warehouses, stables, motor parks, and offices should be accessible by roads. Stables, incinerators, and latrines should be located where prevailing winds will carry annoying odors away from the camp. Latrines should be located as far away from kitchens as practicable to lessen the fly nuisance. A compact layout is preferable, but provision should be made for future possible expansion. Water-bearing fire preventive systems are too expensive to be used; chemical carts are practicable. Figure 102 shows a typical semipermanent camp layout for a triangular infantry division.

b. Rule of thumb.—A rough rule for determining the area of a semipermanent camp for any unit is as follows:

50 square yards per man.

60 square yards per animal.

150 square yards per vehicle.

To use the rule, multiply each unit figure by the corresponding number of men, animals, and vehicles in the unit, and add the products. Application of this rule to Tables of Organization gives the data tabulated below. To this total must be added the area needed for general supply, training, station hospital, railroad yards, etc. The figures also apply approximately to tent camps where space is available for erection without crowding.

■ 183. Barracks.—In the theater of operations, a fair assumption is that barracks will have to be provided for 60 percent of the total force plus 100 percent of the prisoners. In any particular camp, barracks must be provided for all of the troops, and may have to be provided for civilian labor. Barrack space is provided on a basis of 50 men per standard building,

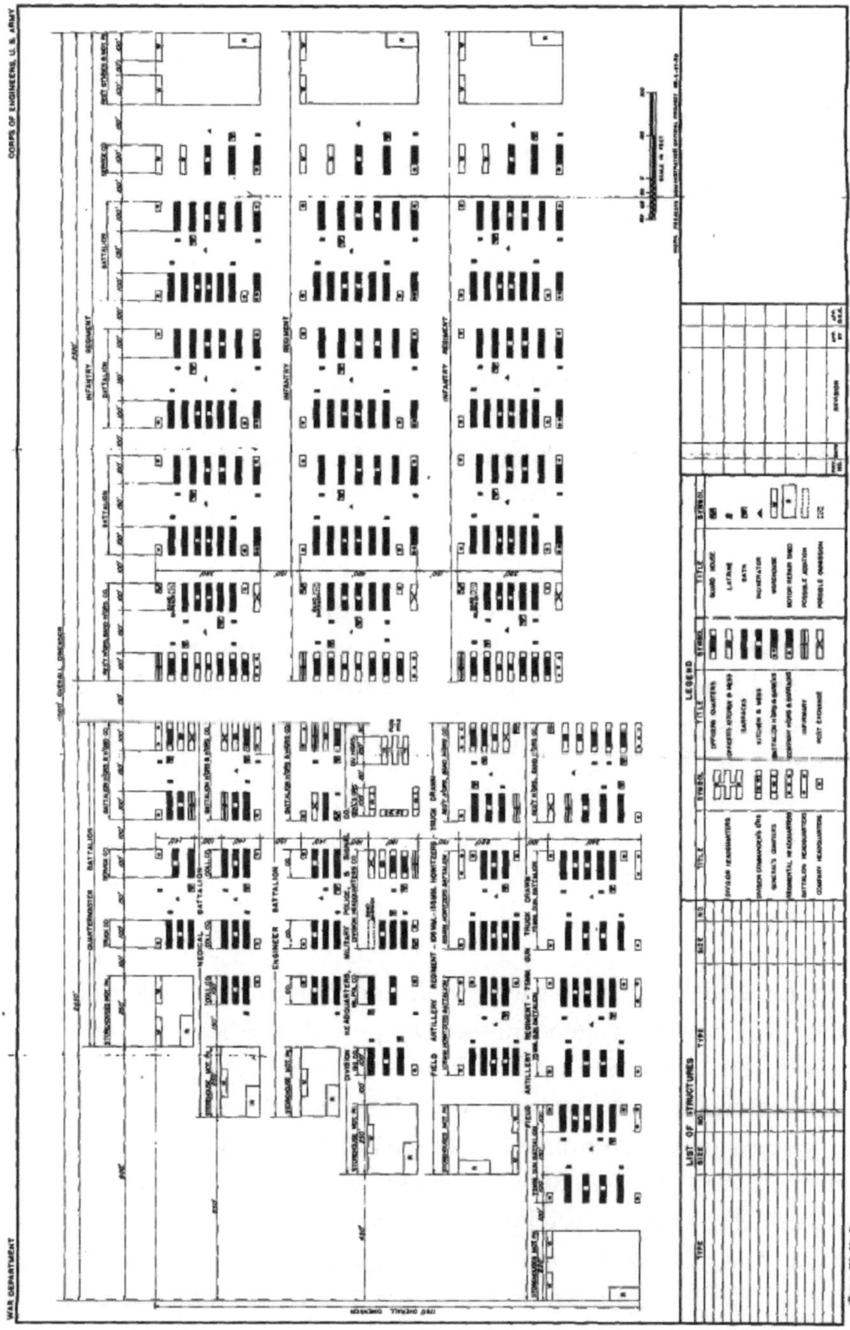

Ⓐ Unit area lay-out.

FIGURE [illegible].—Cantonment, triangular infantry division.

20 by 100 feet. An air space of 400 cubic feet per man is required as a minimum. One hundred men per standard building, 20 by 100 feet, can be sheltered in emergencies, but this is undesirable from a health standpoint. Bunks, of the double-decker type where space is scarce, should be provided for all men. [. . .]

■ 184. OFFICERS' QUARTERS.—The standard building is adapted for officers' quarters by partitioning off double rooms 8 by 16 feet. A messroom and kitchen may be installed at one end.

■ 185. KITCHEN AND MESS HALLS.—The standard building is adapted for use as a mess hall suitable for 120 men. A space 20 by 12 feet at one end suffices for the kitchen. Two buildings 20 by 100 feet can be combined to give a mess hall suitable for 240 men.

■ 186. BATHHOUSES.—A minimum of one bathhouse per battalion area should be provided. This small allowance requires careful administration and supervision in the use of the bathhouses. One bathhouse per company, troop, or battery should be provided wherever possible. A company bathhouse and lavatory, 20 by 24 feet, is constructed in a manner similar to the standard building.

■ 187. LAVATORIES.—Lavatories with ablution and scrubbing benches are installed at the rate of one per company whenever possible. The number of facilities depends on the water available.

■ 188. LATRINES.—Water-bearing sewerage seldom will be used. Simple latrines placed over pits generally will be provided. About one seat to 20 men is desirable, but one seat to 40 men will suffice. A standard 12-seat latrine [can be located] in a building 20 by 8 feet. If necessary, the shelter may be replaced with a simple burlap screen. Where water-bearing sewerage is installed, a simple septic tank may be constructed underground. This consists of a tank served by inlet and outlet pipes, with baffles to slow the flow. The ratio of length to width should be about 4:1, and the depth from 5 to 10 feet. Concrete is a satisfactory material. Organic matter is largely dissolved by bacterial action, and the tank requires cleaning only once or twice a year. The velocity of flow should be under 1 foot per minute. The tank capacity should be about 2 days of sewage. For officers' and nurses' quarters a pail latrine may be used, consisting of a latrine box with removable pails underneath instead of the pit.

■ 189. SHEDS.—*a. Storage and repair shelters.*—An open-sided storage shed [is] suitable for use where supplies are placed under the roof by hand. The roof must be raised by use of longer posts to permit vehicles to be driven underneath. This type of shed is suitable for vehicle repair work and storage of spare parts. All vehicles are stored outside. In this connection, vehicle storage areas should be graded to provide drainage. Roadways serving repair and storage areas should be surfaced, but not the storage areas themselves.

b. Animal shelters.—With mangers and feed racks installed, the open-sided storage shed makes a satisfactory horse shelter. The sides may be sheathed if necessary. [This type of building can include] a center grain-storage room, feed racks, mangers, and covered picket line.

■ 190. LAYOUT OF HOSPITALS.—*a. Percentage allowances.*—Total hospitalization requirements ordinarily vary between 5 and 15 percent of the total strength of the troops in the theater of operations. Prolonged fighting and unhealthful conditions may combine to necessitate total hospital provision for considerably more than 15 percent. Station hospitals at semipermanent camps should provide for hospitalization of about 5 percent of the troops in the area. The remainder of the requirement is provided in general hospitals.

b. Space.—Hospital space allowances necessarily exceed those in barracks. A minimum space at each bed of 60 square feet per patient is required. An additional minimum allowance of 30 to 35 square feet per patient should be provided for administration, supply, operating rooms, and accommodation of hospital personnel. General hospitals are constructed in 1,000-bed units; in emergency their capacity can be increased to 1,200 beds. Station hospitals are constructed in 250-bed units. Figure 114 shows a typical layout for a 250-bed station hospital.

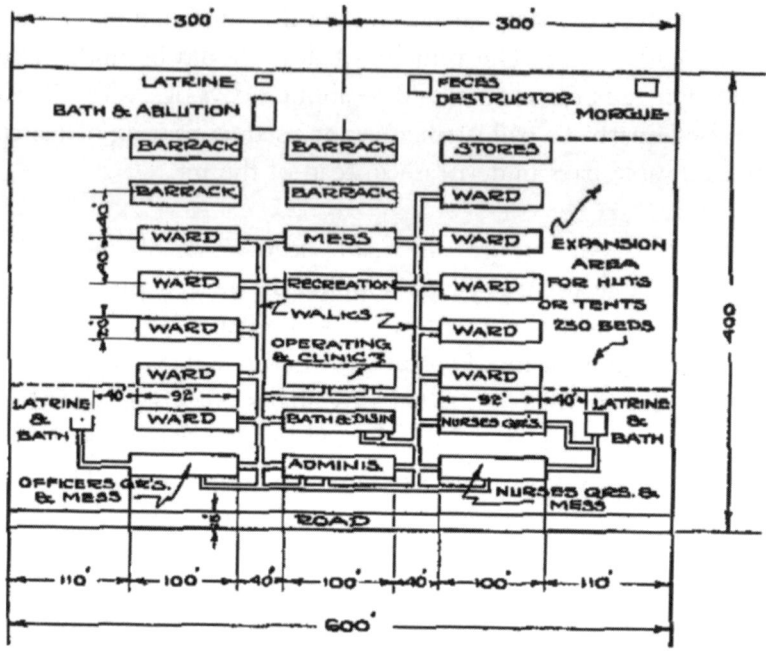

Figure 114. Typical layout for 250-bed station hospital. Individual wards are constructed for normal capacities of 25 and 50 patients.

■ 191. WATER.—Pipe distribution of water is provided for semipermanent camps and hospitals whenever possible. A general figure of 30 gallons per man and 10 gallons per animal per day is reasonable for estimates for semipermanent camps.

■ 192. ELECTRIC POWER.—Electricity is provided to a limited extent in semipermanent camps and hospitals. For camp estimates, allow four 25-watt lights per barrack, 20 by 100 feet, and one 40-watt light per officer.

■ 193. HEATING.—All heating of wartime buildings and tents should be by stoves. Coal, wood, or gasoline may be used as a fuel, depending on which is most plentiful.

CHAPTER 3

Roads, Bridges, and Airfields

To a large degree, the success or failure of any major U.S. Army campaign in World War II depended upon roads and bridges. Offensive operations in particular relied upon a fast and efficient flow of supplies out to frontline forces that, all being well, were pushing further and further away from main supply hubs. For example, in the planning for D-Day and the Normandy campaign it was expected that by D+10 14,700 long tons of supplies were to be discharged from the landing zone beaches and ports every day, increasing to 45,950 long tons by D+90. None of this would be possible if the road and rail network was not free flowing, and it was the engineers' responsibility to ensure that it was.

Much of this effort was dedicated to repairing and optimizing the existing road network and to seizing bridges intact or repairing or replacing those blown up by the retreating enemy. One particularly famous bridge seizure operation was that conducted by soldiers of the Company B, 9th Armored Engineer Battalion (AEB) on March 7, 1945, when they made a dash onto the mighty Ludendorff bridge across the Rhine at Remagen, removing initiating devices from thousands of pounds of live-wired explosives before the Germans could blow the structure. But in many cases, bridges were downed, and the engineers became respected masters at erecting floating ponton or non-floating bridges with exceptional speed, meaning that even the widest river was only a temporary barrier to the advance.

In the following sections from FM 5-10, Engineer Field Manual: Communications, Constructions, and Utilities *(1940) we get a sense of the complexity of challenge behind the construction of road and bridge systems. As we shall see, both roads and bridges required not only a solid grasp of engineering*

The National Bridge over the Penfeld River in France lies in a twisted wreck in September 1944, while U.S. engineers contemplate how to restore traffic across the river. (U.S. Army)

fundamentals such as load bearing and material properties, but also a knowledge of how military traffic moved and flowed in both theory and reality.

★★★

From FM 5-10, *Engineer Field Manual: Communications, Constructions, and Utilities* (1940)

CHAPTER 1
ROADS

SECTION I
GENERAL

[…]

■ 3. Road Capacity.—By this term is meant the number of vehicles which can be moved over the road in a given period of time. Capacity is dependent upon the width, structural character, and conditions of the road, and upon the regulation of traffic. Just as a very poor road may have no capacity at all, so the best one may have none with poor traffic control. A good three-track, two-way road is advantageous in that it permits the passage of columns traveling at different rates of speed in the same direction. A four-track highway is, of course, even more desirable. Structural character and condition influence the number of break-downs, which in turn result in traffic blocks. Good road conditions, therefore, assist in the continuous movement of columns without interruption. Despite a high standard of road construction, the increase in speed of a continuous column will be minimized by the necessary increase in distance between vehicles. From observation it has been found that the maximum capacity is obtained when the speed is about 16 miles per hour, at which speed the distances between vehicles may be reduced to 15 feet. Experience further indicates, however, that increases in speeds up to 25 miles per hour have only a small effect in reducing the maximum capacity. The theoretical capacity of a single lane, under ideal conditions, is about 2,000 vehicles per hour. However, allowing for break-downs and normal delays due to closing up and extending of columns, it is probable that the actual maximum capacity will be only about half of the theoretical. Allowing for the additional delays inherent to vehicles in convoy, it is probable that 750 vehicles per hour is about the working maximum under normal conditions in one-way traffic. If cross-traffic becomes a consideration, as in the case of "turn-out" and "turn-in" traffic on a two-way road, the capacity in each direction may be taken as not to exceed an average of 250 vehicles per hour, with speeds up to 25 miles per hour.

■ 4. TACTICAL REQUIREMENTS.—At least one good two-track road or its equivalent is necessary for each infantry division in line. Corps and army troops should be provided additional roads consistent with their needs in any particular situation. The distance at which troops can be supplied by a road varies with the road conditions, the amount of supplies necessary, the type of transportation available, and the tactical situation.

■ 5. TRAFFIC REGULATION ON ROADS.—In order that roads may give the maximum of service it is necessary to regulate the traffic upon them, both as to direction and speed, and to separate slowly moving columns from those which are able to progress more rapidly. Heavy wheeled vehicles, traveling at high velocities, cause considerable damage at curves and near the edges of a road as a result of shear engendered by the driving wheels in traction. Track-laying vehicles cause even greater relative damage and, when moved at excessive speeds, may so impair the road surface as to make major repairs necessary. For these reasons the engineers should be consulted in the preparation of traffic-control regulations and should be called upon for recommendations as to limiting speeds, directions of movement, and classes of vehicles permitted on various roads. In the case of roads under construction, engineers will direct traffic in accordance with the technical considerations involved and in compliance with instructions by higher authority.

■ 6. NOMENCLATURE.—*a. Axial road.*—A road leading toward the front and generally perpendicular thereto. When designated as the principal traffic artery of a division or higher unit, it is called a main supply road. Generally, main supply roads will not be designated in an independent unit in order to allow maximum flexibility in the use of the roads.

b. Belt road.—A road generally parallel to the front; also known as a lateral road.

c. Course.—A layer of road material parallel to the subgrade.

(1) *Base course.*—The course which rests upon the subgrade or blanket course and supports the top course.

(2) *Blanket (or insulation) course.*—A layer of aggregate, 1 or more inches in thickness, composed of particles well graded from coarse to fine and generally without bituminous binder. Its function is to prevent the underlying materials of a plastic subgrade from migrating upward into the open spaces of a base course composed of relatively large stones.

(3) *Top course (wearing).*—The course last applied to form the finished surface of the road.

d. Crown.—The difference in elevation between the center of the roadway and its edges; it is usually expressed in inches of crown or in inches of rise per foot.

e. Ditch.—The open side drain of a road, designed to carry water running to it from the roadway and adjacent side slopes.

f. Grade.—A line along the center of the road, which, when viewed horizontally, defines the top-surface profile of a longitudinal section of the finished road. Also, the rate of ascent or descent of a road, expressed in percentage or otherwise.

g. Metal.—Broken stone, gravel, slag, or similar material used in road construction or maintenance.

h. Reserved road.—A road reserved by higher authority for designated traffic.

i. Restricted traffic.—Limitations imposed as to the character of traffic, its speeds, loads, hours of moving, etc.

j. Shoulder.—The portion of a roadway between the edge of the metaled wearing course and the ditch.

k. Subgrade.—The upper surface of the natural foundation upon which the blanket or base course is laid.

■ 7. CLASSIFICATION OF MILITARY ROADS.—*a. Standard roadways.*—All roads designed and constructed in general accordance with civil practice may be considered as belonging to this classification, within which they may be further classified as having:

(1) High-type surfaces, composed of paved surfaces of blocks, cement concrete, bituminous concrete, or macadam.

(2) Low-type surfaces of gravel, shale, shell, coral, sand-clay, or earth.

b. Improvised or hasty roads.—Under this heading may be considered all roads of cruder nature, intended by way of expedients to move traffic across otherwise impassable zones or areas. Such roads are similar in type and purpose to the temporary construction roads built by contractors in civil practice. Included are plank roads, corduroy roads, metal mesh roads, tread roads, and trails.

■ 8. MINIMUM DESIGN REQUIREMENTS.—Except for improvised or hasty roads, built to meet the exigencies of a particular situation, all roads should be constructed with a view to their use by heavy motor vehicles and should be designed to facilitate such movements. With this in mind, certain minimum standards are established, though it is realized that failure to achieve these may, in many situations, be the rule rather than the exception. The real criterion is the successful passage of traffic to meet the tactical situation. For roads built in times of peace for possible use in war it is desirable that heavy duty surfacing be provided, but high-type, flexible surfacing will usually be satisfactory. Such surfacing should be at least 20 feet wide and should be capable of supporting 9,000-pound wheel loads on pneumatic tires. This type of road will support any track-laying vehicle in the service. In flat or rolling country, grades should not exceed 5 percent, or curvature, 6°; in mountainous regions these limits may be increased to 8 percent and 14°, respectively. In the case of tactical roads, the surfacing should be as good as conditions permit and should be adequate to support the maximum wheel loads of the unit for which the roads are built or maintained. A minimum width of 9 feet is desirable for a single column of traffic, while 18 feet are required for two columns. Grades should not exceed 10 percent, and sharp curves should be avoided. In general, the radius of curvature should be in excess of 150 feet, and, if this is not possible, additional lanes, 10 feet wide, should be provided. Overhead clearance should be at least 11 feet and 14 feet if possible.

■ 9. MATERIALS.—In order to expedite construction, reconstruction, maintenance, or repair, and to conserve transportation, military road construction utilizes local materials wherever possible. These materials may be earth, gravel, shell, coral, cinders, rock, broken concrete, blocks, bricks, or timber.

■ 10. TRAFFIC SIGNS.—*a. Necessity.*—Traffic signs, now universally used in peace, are much more necessary in war to guide constantly shifting troops and transport in unfamiliar regions. They are used to mark dangerous curves and crossings; geographical locations, such as towns and villages and the important points therein; cross roads and road junctions, indicating where the roads lead therefrom; the location of and direction to important military centers, such as unit headquarters, depots, parks, dumps, refilling and relay points, airdromes, hospitals, ambulance stations, collecting stations, aid stations, artillery positions, assembly positions, distributing points, railheads, regulating stations, and other establishments. Road and traffic signs are placed by or under the supervision of the engineers, in accordance with the road circulation plan.

b. Character.—Traffic signs should be simple and so arranged as to be unmistakable in their meaning. Their size should be commensurate with the purpose for which they are intended. Lettering on signs along a trail used only by foot troops should be from 1½ to 5 inches high whereas, on principal roads, the signs should be provided with letters from 12 to 15 inches high. The letters and their background should be in sharp contrast, such as black against white or chrome yellow. Luminous paint should be used if available. Especially important signs should be illuminated by lanterns or electric lights at night if the situation permits. To catch the eye quickly and aid illiterates, distinguishing marks should be used, such as division or branch insignia where applicable. Directional arrows should be directive and clear.

U.S. engineers of the 167th Combat Engineer Battalion clear roads through Hanover, Germany, a city with widespread destruction from Allied bombing. (Signal Corps Archive)

■ 11. EFFECTS OF TERRAIN AND CLIMATE ON ROADS AND ROAD WORK.—*a. General.*—One of the principal factors controling a choice of road type is that of climate. In dry, desert countries different problems are presented than in arctic wastes or swampy, backwater country; a road well suited to the one condition may, therefore, fail utterly in the other. The most practical way of determining the kind of road best adapted to local conditions is to observe the roads already built and in use. The advice of local construction men and engineers should be sought as to type of road to construct. If no roads are in existence to serve as guides, principal consideration should be given to moisture content and the means of increasing or diminishing it.

b. Roads in arid regions.—These form an exception to the general run of roads, in that moisture is lacking as a binder. It must therefore be supplied, or some means of surfacing must be found which will not disintegrate as it dries. Certain chemicals, such as calcium chloride, serve to collect moisture from the air [. . .]. The same chemicals also serve to reduce or prevent dust, which, when it rises in clouds, may betray the movement of a column. Salt water or sea water, if available locally, may be hauled in tank trucks to advantage, as a substitute for fresh water.

c. Roads in cold climates.—In Arctic regions, where the road surface is permanently frozen, the main consideration may be that of preventing slipping. This may be accomplished by covering the surface with ashes, cinders, sand, pea gravel, or similar materials. When freezing and thawing occur alternately, attention must be given to drainage of the subgrade in order to prevent heaving. Snow and ice may be removed by hand shoveling or by the use of special machinery.

d. Roads in tropics.—Because of the abundance of rain during certain seasons of the year, drainage is an important consideration. In general there will be a lack of material for concrete or gravel roads, but use can be made of lava rock or coral to construct very suitable roads. The latter is quite soft, hence roads built with it are subject to considerable wear and must be carefully maintained.

■ 12. IMPROVED AND UNIMPROVED ROADS.—Although these words have no clear-cut technical significance, they are often used in a tactical sense to differentiate between all-weather, hard-surfaced roads, suitable for motor traffic, and low-grade roads adaptable only to the movement of animal-drawn vehicles and foot troops, except in the best of weather. In general an improved road may be considered as a high-type road with two or more traffic lanes, while an unimproved one may be counted upon to require considerable improvement and maintenance and to afford only one lane for the movement of traffic.

[. . .]

CHAPTER 2
BRIDGING, GENERAL

Section I
BRIDGING CONSIDERATIONS

■ 46. ROLE OF BRIDGES.—*a.* A line of communications, as defined by a road or railroad, must be held intact throughout its entire length. Breaks in the line mean breaks in the system of supply and, as in a chain, the whole can be no stronger than its weakest link. Bridges afford a means of carrying the line forward across streams, rivers, gulches, ravines, or draws, which otherwise would constitute serious obstacles to the rapid movement of troops, supplies, or armaments. By their very nature they are extremely vulnerable, hence it is of first importance to protect and maintain all crossings which are subject to the present or future use of friendly troops. The need for new construction must be carefully weighed when laying out a system of communications, for if such work is excessive a relocation of the route may be indicated. Under certain circumstances it may be highly desirable to circumvent a terrain depression at the cost of a longer road to build and maintain; on the other hand, it may be possible to effect a material saving in time and labor by the construction of a suitable crossing. Consideration should always be given to the possibility of substituting for bridges low embankments with culverts, since the latter are less vulnerable to hostile aerial attacks and, in cases where relocations are possible, may be less costly in materials of construction. When the need for a bridge is definitely indicated, decisions must be made as to the type necessary to carry the anticipated loads, the permanence required in the structure, and the materials and labor available for its construction. As in the case of roads, new construction should be held to a minimum, full advantage being taken of existing structures. Also, full use must be made of locally available materials, to which end reconnaissances must be extensive and continuous.

b. Necessity for the utmost speed in bridging operations requires that every possible preparation be made for the construction of bridges that will be required in a particular operation. Information regarding spans of bridges required for the advance is necessary in order that the equipment will be suitable as to type and amount. The material required and the engineer troops necessary for the construction should be well forward in order to expedite the erection of the bridges.

[. . .]

■ 48. BRIDGE NOMENCLATURE.—*a. General.*—The essential, basic components of a bridge are the substructure and the superstructure. Included in the substructure, in addition to the abutments and foundations, are the supports upon which the superstructure is carried. These latter may take the form of piers, bents, or pontons. The superstructure constitutes the remaining upper part of the structure, including the stringers, flooring, stiffeners, and overhead supports.

b. Definitions.—Some of the more common, general words used in connection with the subject of military bridging may be defined as follows:

(1) *Abutment.*—The ground support of the superstructure at an end of the bridge.

(2) *Balk.*—The standardized stringers of a floating bridge.

(3) *Bent.*—An intermediate, transverse support consisting of a framework of horizontal and vertical members, usually requiring external bracing for stability. (Short pile bents do not require longitudinal bracing.)

(4) *Chess.*—The standardized floor-planks of a floating bridge.

(5) *Flooring.*—The deck covering which forms a roadway for traffic across a bridge.

(6) *Footing.*—The arrangement whereby loads from supports or bridge seats are distributed over a greater ground area as a means of reducing unit pressures.

(7) *Girder.*—A simple or built-up beam, usually of steel, designed to carry floor loads to piers or abutments.

(8) *Pier.*—An intermediate support of masonry, reinforced concrete, cribwork, or of several bents so constructed as to form an integral unit needing no additional bracing for stability.

(9) *Ponton.*—A float, often in the form of a boat, used to provide buoyance for the superstructure and imposed loads of a bridge.

(10) *Sill.*—The member of a support which rests directly upon the ground or a footing.

(11) *Span.*—The portion of a bridge between centers of two adjacent supports; alternately, the distance between such centers.

(12) *Stiffener.*—A girder, or truss, used to stiffen the superstructure and to aid in carrying the weights imposed upon it.

(13) *Stringer.*—One of a number of longitudinal members resting upon end supports and carrying the flooring.

(14) *Trestle.*—Same as a bent; often referred to as "trestle bent."

(15) *Truss.*—A compound beam, the parts of which are arranged to form one or more triangles in the same plane so that the beam will transmit roadway loads from the floor system to the abutments or to intermediate vertical supports called panel points.

■ 49. TYPES OF BRIDGES.—Bridges may be classified according to inherent fixity (as whether they are fixed, movable, or portable); according to the number of spans (single or multiple); according to the type of supports employed; or according to the materials used in construction. Other classifications may be as to the magnitude and character of loads to be carried, and as to the general character of structure; that is, whether it is a bascule, cantilever, suspension, or ponton bridge. But even these are not all-inclusive methods of classification; the breadth of subject is such that no clear-cut cataloging and indexing are possible. Considering the matter, however, from a standpoint of what bridges are adaptable to military use, logical classifications can be made with regard to loads carried and types of support.

a. *Loads carried.*

(1) Foot and cart bridges.

(2) Light vehicular bridges.

(3) Heavy vehicular bridges.

b. Types of support.—It is possible to group together all fixed supports, such as cribs, piers, bents (and abutments only, in the case of simple stringer bridges) and arrive at the classification:

(1) Floating bridges.

(2) Nonfloating bridges.

The following table shows the more common types of military bridges arranged according to these classifications. Omitted from the table are such types normal to civil practice as suspension bridges, steel cantilevers, movable (lift) span bridges, railroad bridges, etc. [. . .]

Types of military bridges

Type	Classification	
	Floating	Non–floating
Foot and cart	Standard footbridge equipment (model 1935) (duckboards on floats). Kapok footbridge (obsolescent) (crates filled with kapok pillows). Improvised types using gasoline drums, cans, wooden floats, or other expedients.	Simple stringer. Light trestle bents: Pile. Framed. Lashed spars.
Light vehicular	Metal pontons (model 1926, 7½-ton). Metal pontons (model 1938, 10-ton). Pontons with canvas-covered wood frames (model 1869, light ponton) (obsolescent). Wooden pontons (model 1869, heavy pontons) (obsolescent). Improvised raft supports of wood or metal drums, logs, boats, etc.	Simple stringer (wood or steel). Framed trestle bents. Framed trestle piers. Pile trestle bents. Pile trestle piers. Lashed spar trestles. Portable truss girders (H–10 loading).

Heavy vehicular	Heavy ponton equipage (model 1924, 23-ton). Light (metal) ponton equipage (model 1926, 7½-ton) reinforced to carry 15 tons. Light (metal) ponton equipage (model 1938, 10-ton) reinforced to carry 20 tons. Improvised raft supports.	Demountable steel truss girder (23-ton capacity) (obsolete). Portable truss girders (H-20 loading). Simple stringers. Framed trestle bents. Framed trestle piers. Pile trestle bents. Pile trestle piers. Crib piers (wood or steel). Concrete or masonry piers.

■ 50. MATERIALS OF CONSTRUCTION.—*a. Wood.*—Except for the portable ponton and steel truss girder bridges adopted as standard in the Army, most bridges will be constructed of timbers in time of war. For this reason it is desirable that the military engineer have definite knowledge of the commercial sizes in which lumber and timber are generally obtainable.

[. . .]

Ordinary lengths are 12 to 20 feet, lengths of 20 to 26 feet are less common, and 27- to 32-foot lengths are obtained with difficulty. In ordering lumber, bills of material are prepared to show for each item the number of pieces required, and the cross section, length, grade, and surfacing of each piece. Four pieces of No. 1 common Douglas fir, each 16 feet long, 12 inches wide, and 6 inches thick, surfaced on all four sides, would be designated:

4–6" × 12" × 16', Douglas fir, No. 1 common, S4S

Lumber is usually sold by board feet (*FBM*=feet board measure), the unit being a board 1 foot long, 1 foot wide, and 1 inch thick. To compute the FBM in any rectangular timber the length in feet is multiplied by the width in feet and by the thickness in inches. Thus, in the four pieces of Douglas fir mentioned, there would be:

4×6×1×16=384 *FBM*.

In general, the cross section of surfaced pieces will be somewhat less than unsurfaced pieces. Since the above dimensions are based on unsurfaced pieces, the designer should take into account the decreased area of the surfaced sections.

b. Steel.—When the span is greater than 15 feet, steel stringers are frequently used. These should be designated by standard sections, which in the case of I-beams are represented by depths of 3, 4, 5, 6, 7, 8, 10, 12, 15, 18, 20, and 24 inches and by different weights as given in manufacturers' handbooks. For the standard type, H-15, trestle bridge, suitable *CB* (car building) sections are to be stocked in depots.

A floating footbridge over the Roer River.

■ 51. FACTORS OF SAFETY.—*a. General.*—If a structure were loaded to its ultimate safe stress, the application of any additional load would result in its failure or damage. Although, in theory at least, the structure would continue to stand undamaged until its ultimate safe stress had been exceeded, no such narrow margins are permissible in practice, for the failure or damage of structures may be attended by serious consequences, and in no case is it possible to determine exactly the ultimate strength within the elastic limit of a material or the exact loads to be imposed. Furthermore, perfect construction is not possible. The factor of safety of a material or a structure may be defined as the ratio of its ultimate safe stress, within the elastic limit, to its actual working stress. In civil practice, factors of about 2.2 and 4.0 are applied to steel and wood structures, respectively. Since safety is of relatively less importance in war than in peace and economy of materials is very important, these factors may be reduced to 1.75 for steel and 3.0 for wood in certain cases, particularly in connection with bridges in forward areas or which will be used for only a short time. Under exceptional circumstances, the factors may be reduced still more, but in such cases the bridge should be strengthened as soon as conditions permit. For bridges in the communications zone and large and important bridges in the combat zone which will be used for a considerable time, the civil factors should be used.

b. Impact.—When a load is moved suddenly upon a bridge, exceptional stresses are induced, for which, in civil practice, due allowances are made during design. These are the exceptional and momentary stresses produced by excessive strain before the material has adjusted itself to the load, and the impact produced by the momentum of the load. The former is taken care of in the factor of safety; the latter is taken care of in civil practice by an allowance therefor. In the design of military timber bridges, however, no allowance is ordinarily made for impact; and for steel bridges, including wooden bridges with steel stringers, the factor used is half that of the American Association of State Highway Officials. The formula as modified in the latter case is:

$$I = \frac{25}{(L + 125)}$$

in which I is the percentage of the live load to allow for impact and L is the length in feet of the span considered. It will be seen from the formula that a factor of 20 percent is adequate in any case. Theory must be tempered with both judgment and experience when figures less than 15 percent are obtained or when loads of exceptional character are anticipated.

SECTION II
DESIGN OF BRIDGES

■ 52. APPROACHES.—*a. General.*—The following considerations are important in connection with the siting of bridge approaches:

(1) It should be borne in mind that work on the approaches will often, if not usually, be greater than the work on the bridge itself. The location of the bridge will therefore usually be very dependent on possible approaches and their condition.

(2) The center line of the road from each end of the bridge should continue straight for at least 50 yards in a prolongation of the bridge axis.

(3) Approaches should have a slight upgrade toward the bridge.

(4) Maximum use is to be made of existing roads.

(5) Approach roads should be made wide, with turn-outs or parking areas at each end of the bridge when the necessity for same is indicated.

b. Preparation of the roadway.—If the approach roads are on very bad ground and there is insufficient metal to provide a relatively hard surface up to the abutment, it may be necessary to provide an extra timber span, or to lay a plank road upon the ground surface. Corduroying is also effective in such cases, but less desirable because of the tendency of traffic to side-slip upon it and because of its rough nature. When the approach is very firm, it may not be necessary to use any form of timber decking, but if the roadway has been excavated behind the end dam or if it must be built up, great care must be taken with the backfill, which should be of rock or crushed aggregate, tamped in layers about 3 inches thick. The road surface at the end of the bridge should be of broken stone or similar material laid as macadam unless a more rigid surfacing is possible. To avoid shocks and possible displacement of the bridge due to vehicles striking against its end, the road surface should be built up to about an inch above the flooring. Too high or too low a road surface at this point of juncture may decrease the life of a bridge as much as 50 percent. [...]

c. Maintenance.—Having taken care of the approaches, it is next necessary to provide for maintenance. The tendency of all heavy traffic is to develop holes in the road pavement about 2 feet from the end of the bridge. As explained, this results in a jolting of vehicles, with an attendant increase of impact on the bridge, both vertically and horizontally. With horizontal jars transmitted to the abutment each time a truck passes, serious consequences may quickly ensue. For this reason periodic inspections should be scheduled and continuous maintenance arranged if necessary. Road materials, stacked in convenient piles at each end of the bridge, facilitate the work of repair.

■ 53. ABUTMENTS.—*a. General.*—Abutments for military bridges vary in character according to the planned permanence of the structure. Designs are based on expediency, with primary consideration usually to speed and ease of construction. Although the loads on a semipermanent military bridge may exceed normal highway loads, it is improbable that time will permit the same care of construction in the first case as in the second. Lack of time must be compensated for by increased judgment. The importance of time-saving in connection with abutment construction may be appreciated by consideration of the fact that on bridges of short span—say up to 100 feet—the construction of abutments, similarly to work on the approaches, may require as much time as the entire remainder of the work. In permanent highway construction the abutment is usually constructed of masonry or concrete. In military bridging this type of construction requires too much time. Furthermore, the military bridge is not usually designed to last long periods of time. Hence timber is the material generally used in military abutments.

b. Component parts.—A bridge abutment consists of three distinct parts:

(1) A bank seat which supports the bridge span.

(2) The connection between the bridge and approach.

(3) The retaining walls or other arrangement provided to prevent the earth from sloughing off beneath the bank seat.

c. Bank seat.—The bank seat consists usually of a bridge seat, or timber sill, upon which the end of the span rests, and a timber footing, or mudsill, which serves to distribute the load over a greater area. For heavy steel bridges it is frequently necessary to place a bearing plate between the bridge span and the seat in order to distribute the load over a sufficiently large area to prevent crushing the wooden sill.

d. Construction of footings.—The following principles should be observed in placing timber footings:

(1) Set the footing course about 2 inches higher than its final desired position to allow for settlement.

(2) Do not dig too deeply into the bank. If this is done by mischance, raise the seat by planking rather than by backfilling with earth.

(3) Keep behind the natural slope of the earth, which is usually about 1½ horizontal to 1 vertical. Lacking laboratory facilities, this can be determined for practical purposes by observing old cuts and fills in the immediate vicinity.

(4) Place the pieces of the lower footing course parallel to the axis of the bridge.

(5) Place the bridge seat so that the load comes on the middle of the footing course.

(6) Pill and tamp the footing thoroughly if it is placed in a trench. Provide for drainage if it is placed on a shelf excavation.

(7) The projection area of the footing must be such that the safe bearing pressure of the ground is not exceeded.

(8) The timber must have sufficient cross section to withstand crushing and bending.

e. End dams.—After the bank seat has been built up, and the bridge is in place, an end dam of planks should be placed across the ends of the stringers, to which it may be spiked. The purpose of this is to keep the roadway from caving in between the stringers at the abutment.

f. Retaining walls and revetments.—(1) The object of retaining walls is not only to prevent the earth from sloughing away beneath the bank seat but also to facilitate a shortening of the span. [. . .] When the water is likely to rise above the base of a crib, rock ballast should be supplied, both inside and out.

(2) Revetments are frequently employed to prevent undue scour around or beneath the bank seats in times of flood when velocities are high. Such revetments may take the form of brush mats or riprap, extending from low-water level to the top of banks. If riprap is used, it is desirable to lay it on a 6-inch gravel blanket, providing time and materials are available and the bridge is to be used over a considerable period. Concrete paving is entirely satisfactory as a means of protection, but requires more time to place, and for that reason is seldom used on military bridges. When used, it should be provided with weep holes for the outward flow of water from the saturated bank after recession of flood stages. Surface drains should be provided as a means of preventing erosion around the ends of the bank seats.

U.S. Army engineers construct a rail spur running across a Normandy beach, to facilitate the linkup between offshore vessels and the beachhead, June 1944. (U.S. Navy)

■ 54. DESIGN OF FOUNDATIONS AND FOOTINGS.—*a*. Add the total of live and dead loads on the bridge seat to determine the pressure beneath it, in pounds, allowing 25 percent for impact. Example: 35,000 (dead load) + 10,000 (live load) = 45,000 pounds (total).

b. Divide this figure by the unit bearing power of the soil, determined by inspection, from the following:

Soil	Unit bearing power (pounds per square foot)
Loam	1,500
Silt or clay fills, after 6 months (general)	1,500
Soft clay	2,000
Firm clay, confined sand, or gravel	8,000
Rock	20,000–40,000

Example:

$$\frac{45,000}{1,500} = 30 \text{ sq. ft.}$$

The result is the required bearing area of the footings, or of the bridge seat (sill) if no footings are required. Where footings are required, it should be assumed that the bridge seat (sill) will not provide any bearing and that the footings alone must furnish the full surface area.

[. . .]

■ 56. DEMOLITION CHARGES.—When tactical requirements warrant, demolition chambers should be provided in military bridges beneath the abutment bank seats. Such chambers should be placed at least 5 feet below the bank seats but not below flood level. If this is not possible with a chamber of 10 cubic feet normal capacity, two chambers should be placed, each with a capacity of 6 to 8 cubic feet. Unless a vertical shaft can be sunk quickly and easily from the roadway, the chamber

Engineers sweep for mines at the head of the U.S. advance through Tunisia near Kasserine in February 1943. (Signal Corps Archive)

(or chambers) should be made by excavating diagonally downward from in front of the footings. Both the chamber and shaft should be lined with timber, and 10 feet of tamping should be provided. The whole system should be planned so that the opening to the chamber is accessible at all times and in no danger of being flooded.

■ 57. BRIDGE LOADS.—*a. General.*—An understanding of stresses and their application is essential as an approach to the design of any frame structure. In the case of a bridge, certain stresses are induced by the live traffic loads for which it is designed, while others result from the dead weight of the structure itself. The distinction between live and dead loads is important, because the former may produce impact stresses as discussed in paragraph 51b; also, live loads may be either uniformly distributed along the length or the span or concentrated at one or more places. A given load concentrated and applied at the center of a span will produce twice the bending stress that would be developed by the same load uniformly distributed along the span.

  b. *Permissible military loads.*—(1) The following limitations on gross weights of military vehicles are extracted from AR 850-15:

  (a) The gross weight of a vehicle is defined as the chassis weight, plus the weight of the cab and the entire body, fully equipped and serviced for operation, plus the maximum allowable payload over good roads, and the weight of all operating personnel.

  (b) The gross weight of any vehicle or combination of vehicles designed to accompany the military forces in the field and for movement by highway, or the gross weight carried by any group of two or more axles of such vehicles or combination of vehicles, will not exceed that given by the formula—

$$W=C(L+40)$$

where W=the gross weight of a vehicle or combination of vehicles or the gross weight carried by any group of two or more axles.
     C=a coefficient which is prescribed below for different classes of vehicles.
     L=the distance in feet between the centers of the first and last axles of the vehicle or combination of vehicles or between the centers of the first and last axles of any group of two or more axles.

  (c) In applying the gross weight formula to a track-laying vehicle or to a combination in which there is a track-laying vehicle, the gross weight on the tracks will be considered as two axle loads, each equal to half the gross weight, applied either at the ends of the ground contact length of the tracks or at points halfway between the ends of the ground contact length and the center of the ground contact length, for the purpose of determining the length L, the greater gross weight being allowed in any case.

(*d*) In applying the gross-weight formula the following values of C will be used:

> Type of vehicle   Value of C.
> 1. Vehicles designed to accompany an infantry or cavalry division
> or cavalry mechanized unit in the field, when the distance between
> the first and last axles of the group of axles considered is—
>   18 feet or less............................★325
>   Greater than 18 feet...................★375
> 2. Vehicles designed to accompany an army in the field and for
> movement by highway, when the distance between the first and
> last axles of the group of axles considered is—
>   18 feet or less.............................600
>   Greater than 18 feet.....................700

★This factor is for limiting loads for the present 71-ton ponton bridge (model of 1926). It necessarily will be changed when tests have been conducted with the Standard Portable Highway Bridge (H–10), recently standardized; the contemplated 10-ton ponton bridge will also necessitate a like change in this factor.

(2) Limitations on axle, wheel, and track loads are as follows:

(*a*) The wheels of all vehicles except those operated at not to exceed 10 miles per hour will be equipped with pneumatic tires.

(*b*) No wheel equipped with a low pressure pneumatic tire will carry a load in excess of 9,000 pounds, nor will the total load carried by any axle having wheels equipped with such tires exceed 18,000 pounds. No wheel equipped with high pressure pneumatic, solid, or cushion tires will carry a load in excess of 8,000 pounds, nor will the total load carried by vehicle per lineal foot of ground contact will not exceed 5,000 pounds. An axle load will be the total load on all wheels whose centers are included between two parallel transverse planes 40 inches apart.

(*c*) Track-laying vehicles will have tracks of such length and width that at zero submergence the gross weight of the vehicle per lineal foot of ground contact will not exceed 5,000 pounds and the ground contact pressure will not exceed 15 pounds per square inch.

(*d*) No wheel equipped with pneumatic, solid, or cushion tires will carry a load in excess of 600 pounds for each inch of tire width. The width of pneumatic tires will be taken as the manufacturers' rating. The width of solid rubber and cushion tires will be measured at the flange of the rim.

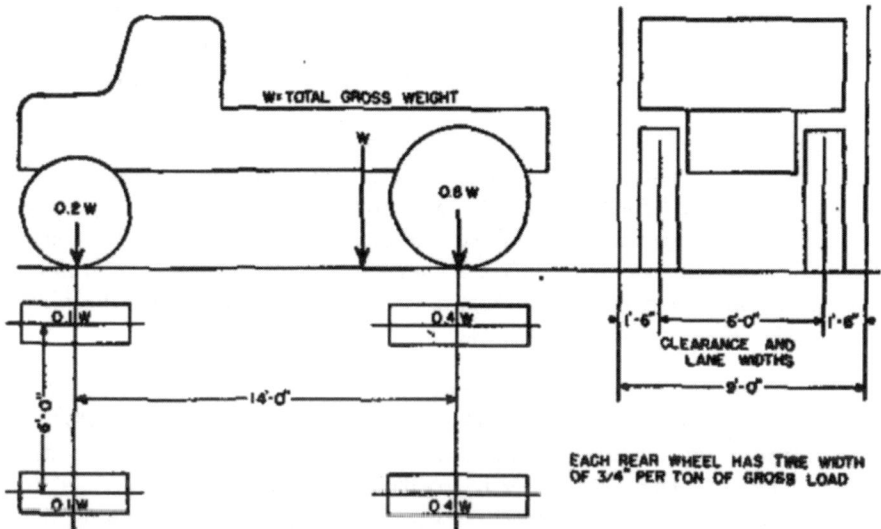

FIGURE 30.—Distribution of wheel loads for design.

Combat Experience and Innovation

From 1944, the U.S. War Department started to publish a regular series of publications entitled Combat Lessons. *The purpose of the series was to "to give to our officers and enlisted men the benefit of the battle experiences of others. To be of maximum benefit these lessons must be disseminated without delay. They do not necessarily represent the carefully considered views of the War Department; they do, however, reflect the actual experiences of combat and, therefore, merit careful reading."*

Combat Lessons *included tactical insights from all branches of the U.S. land forces, including the engineers. The publication was, by 1944, able to draw on first-hand accounts from a broad spectrum of theaters and battles. The following text is taken from Issue No. 7, specifically from 'Engineering Ingenuities', an article collecting the pragmatic lessons of U.S. Army engineers in action. This section focuses principally on the business of constructing bridges and cording roads:*

Supply—in Spite of High Water!

Reported by Staff Sergeants Lewis E. McKenzie and James R. Gaveske, 135th Infantry, ITALY: "On the first crossing of the Volturno River, we came up after dark to the regimental CP and found that there was no way to cross the river by truck. Rations, water, and ammunition had to be gotten across.

"First, we tried a jeep, but the jeep went downstream. The next thing we tried was pulling a 2½- ton 6 × 6 truck across by its winch. This worked okay so we loaded a jeep and supplies on the truck and then dragged it across. We kept that up until we had 4 jeeps across with which we could haul the supplies out to the companies about a mile away. We moved all of the rations and ammunition that night by the same process. We also took all of the casualties back that way."

Bridge Construction Expedient

The successful use of an expedient bridge near Soputa, NEW GUINEA, is reported by *Captain Ralph E. Reed, Company Commander, Combat Engineer Battalion*: "The Girua River is a shallow, sandy-bottomed stream subject to flash floods during the rainy season. These factors made impossible the construction of a standard trestle bridge. Equipment was not available for either a ponton trestle or a pile bent bridge. [Our] method [...] was used to provide the needed bearing for trestle bents. It also prevented scour that would have occurred with standard trestle erection. The entire construction was from round timber.

"Small piling, 4 to 6 inches in diameter, were driven under each sill to form a series of 8 to 10 X-shaped supports. The crossed piling were so driven that the crossing points were at a depth which permitted the sills to bear on the stream bed when fitted into the upper angle. Empty 25-pounder shell cases were used on piling tips to prevent damage from sledging. The upper arms of the piling were wired together and windlassed. Directly upstream from the trestles, holdfasts were driven. The erected bents were anchored by cable to these holdfasts.

"This bridge crossed the stream at a width of 200 feet. It withstood all types of Class 5 traffic and was subjected to many severe floods but was still in good condition when replaced after 6 months."

To the Engineer in the Woods

A *VI Corps* report recommends: "Tank dozers are the most effective means of removing abatis and other log obstacles commonly encountered in wooded areas.

"In the soft loam generally found in wooded terrain, corduroy roads will accommodate heavier traffic and require less maintenance than any other type of improvised surface."

Cording Roads

Sergeant Norman I. Roenbaugh, Infantry, SOUTH-WEST PACIFIC: "The only road that seemed to cope with the heavy mud in the Central Solomons was the heavy-timber corduroy type. The engineers used expedient sills made of 18- to 20-in, logs for stringers."

Preventing Bent Truck Frames

A helpful hint for road-building engineers comes from *Captain Clyde G. Grant, Northwest Service Command*, NORTHWEST CANADA: "When using the 1½-ton dump truck for extensive road-surfacing operations, it was found that the frame bent at a point between the rear of the cab and the front of the dump bed. The condition was corrected by welding a sheet of 5- by 6-inch steel plate along the frame on each side."

Getting Cable Across Streams

Commanding Officer, Reserve Combat Command, 6th Armored Division, FRANCE: "Cables can be carried over streams by boat or shot over by grenade.

"When using the boat method over swift streams, the first boat should be a light one with a small outboard motor, carrying two or three men and a light line for drawing the cable. Large boats are harder to handle before the cable is in, and the larger propellers are likely to strike obstacles and become disabled.

"On one occasion when two boats had been lost while attempting to carry a cable over a swift stream, we used a rifle grenade to do the job. Engineer tape was tied to the grenade and fired across the river. Men on the far shore then pulled over a telephone wire and, finally, the cable."

★★★

The following text from FM 21-105, Basic Field Manual: Engineer Soldier's Handbook *(1943), partly expands upon the theme of bridge construction*

Members of B Company, 208th Engineer Combat Battalion, use a 2½-ton truck to push logs into a Dutch sawmill in December 1944. (U.S. Army)

outlined above, but also brings in the topic of airfield construction. The rapid establishment of forward airfields became critical in all theaters of U.S. operations, as such bases enabled tactical support aircraft to keep the frontline of advance within their operating radius and, in the Pacific theater in particular, strategic bombers to bring the Japanese homelands to within striking distance. While the engineers laid dozens of airfields in asphalt and concrete, these required heavy equipment and more time-consuming processes. If time was of the essence, use of prefabricated runway structures such as the Pierced Steel Planking (PSP), aka "Marston Mats," could rapidly expedite matters. Just a handful of committed engineers could lay a runway 200ft (61m) wide and 5,000ft (1,500m) long in only two days.

★★★

From FM 21-105, *Basic Field Manual: Engineer Soldier's Handbook* (1943)

CHAPTER 9
BRIDGES

■ 67. GENERAL.—An unfordable river is a difficult obstacle to an advancing army. The enemy, therefore, destroys all possible bridges in the path of advance. It is the job of engineers to build substitute bridges in the shortest possible time.

■ 68. SPEED AND TEAMWORK.—Army engineers should be the fastest bridge builders in the world. They can build fixed and floating bridges quickly because of two things:

a. Their equipment is designed for hasty, rugged construction.

b. Their building crews are trained in teamwork and speed.

The second factor depends on the individual engineer soldier. Hundreds of feet of bridge must be built in a few hours under difficult conditions; you will be very tired; sometimes you will be under enemy fire; you may be cold and wet; or hot and dry. But upon you depends so much that you must overcome all handicaps. An army may be waiting for the products of your toil. You must give all you have in you to do the job *on time!*

[. . .]

■ 70. FIXED BRIDGES.—The most common military fixed bridges are the simple stringer bridge, the trestle-bent bridge, the light portable steel bridge, often called H–10, and the Bailey bridge.

a. The simple stringer bridge is usually short. It consists of three elementary parts: two abutments, a single span of stringers, and a floor. Two types of abutments are used, one for use with soft approach roadways, the other for use with firm roadways. Every engineer soldier should know how to construct a simple stringer bridge. Trestle bridges are merely a succession of simple spans in which the trestles take the place of abutments. Timber stringers are seldom used in spans of over 15 feet or steel stringers in spans of over 25 feet.

b. *The trestle-bent bridge* consists of two or more stringer spans. The supports between the abutments are trestle bents.

c. *The light portable steel bridge* (H–10) consists of two trusses (assembled by manpower, in lengths up to 72 feet) supporting a one-track timber deck. The 12-foot girder sections are carried in trucks and bolted together to build the bridge. The deck planks are held in place by siderail clamps, which hold the siderails to the trusses.

d. *The Bailey bridge* is an English panel bridge built to carry heavy loads. It can carry 70 tons on spans up to 120 feet, but requires time for erection for these loads.

■ 71. FLOATING BRIDGES.—There are a number of different kinds of military floating bridges in use. They are carried by different kinds of engineer units.

a. The *footbridge* M1938 is constructed of separate rafts called "bays," each 12 feet long, consisting of a duckboard supported by two floats.

b. The *light ponton bridge* M1938 is a floating bridge capable of carrying 10-ton traffic in one direction.

c. The *heavy ponton bridge* M1940, 25-ton, is similar to the light ponton bridge but is much heavier and will carry 25 tons with normal construction. It can be reinforced to carry 35 tons.

d. The *steel treadway bridge* is designed to carry medium tanks. It has steel treadways for runways, which are emplaced by means of a truck-mounted crane. It uses special rubber pontons.

e. The *pneumatic bridge* M3, made with 12-ton floats, can carry 13 tons, or, when reinforced, 18 tons. It uses regular 10-ton ponton balk and chess for the floor system.

[. . .]

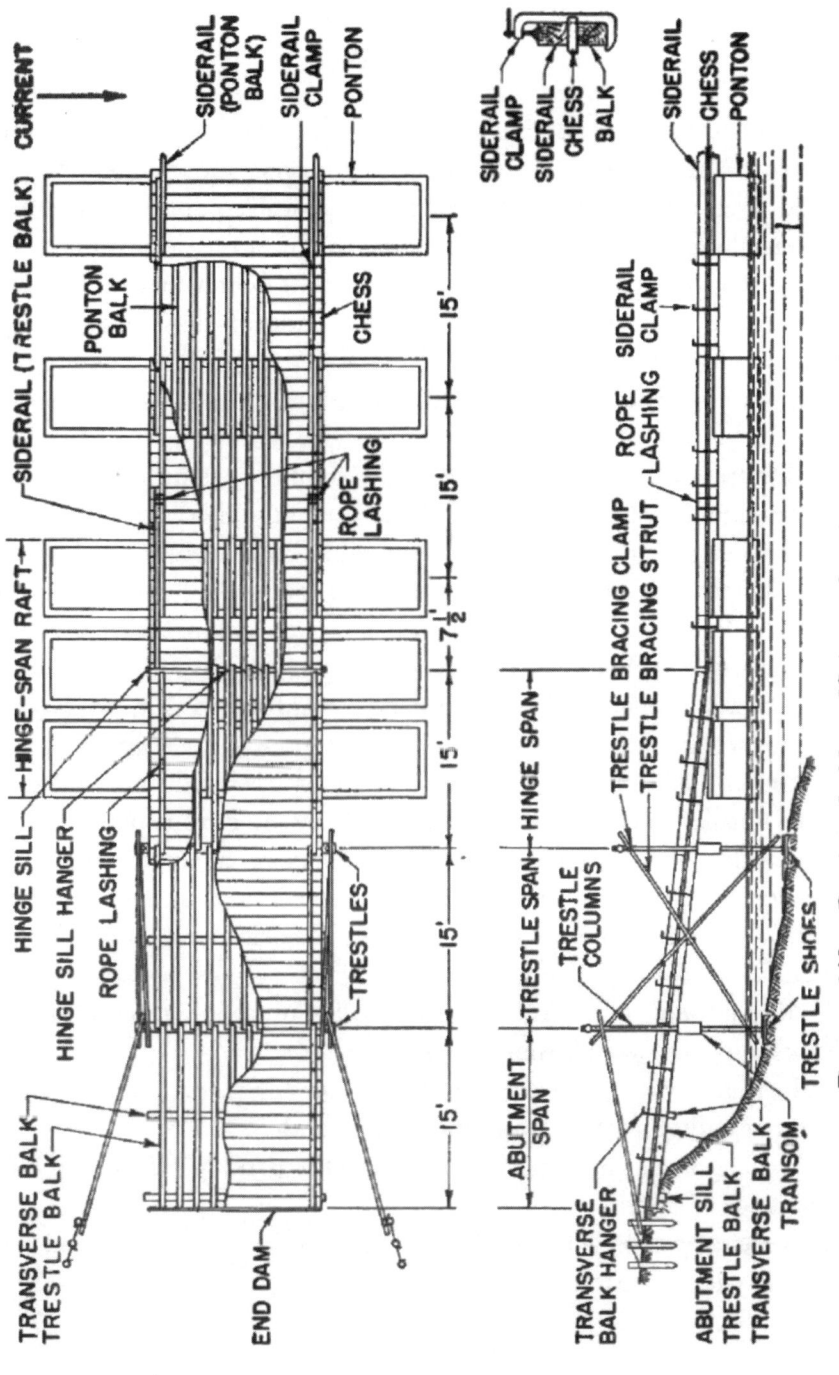

FIGURE 119.—Ponton bridge, 25-ton, showing names of parts.

CHAPTER 11
AIRDROMES

■ 78. ENGINEERS AND THE FLIERS.—Engineers prepare the way for the Army Air Forces. They are builders, defenders, and maintainers of airfields. Well armed and with much mechanical equipment, aviation engineers plunge into the wildest country, the most forward battle areas, and build swiftly the bases from which our aircraft fight. Once built, these bases must be defended and kept in good condition despite bombing, strafing, or artillery bombardment. The flier depends upon the engineer for this support. You must not let him down.

■ 79. DEFINITIONS.—All army engineers should be familiar with the general design, construction, and nomenclature of military airfields, since any general engineer troops may be required to build them. Airfield itself is the general term applied to any area used for landing and taking-off of aircraft. Following are some of the more common terms used in connection with airfields.

a. Advanced landing field.—Temporary airfield near front, with only minimum servicing facilities.

b. Air base.—An area including a parent or base airdrome and one or more smaller airdromes situated at some distance from parent field. Smaller airdromes are sometimes called auxiliary or satellite fields. They depend on the base airdrome for complete repair and supply facilities.

c. Airdrome.—Landing field with facilities for shelter, supply, and repair of aircraft.

d. Alternate airdrome.—Airfield available for use of air force units, in addition to one to which they are assigned.

e. Approach zone.—Cleared area, which allows friendly aircraft to see the field at a distance and come in at a low glide.

f. Apron.—Surfaced or paved area used for parking, servicing, and maintenance of aircraft.

g. Dispersal parking area.—Area in vicinity of airdrome, used for dispersed (widely separated) parking of aircraft.

h. Dispersed airdrome.—Airdrome in which runways, technical facilities, and housing are spread out to aid concealment and lessen damage in event of a bomb hit.

i. Field airdrome.—Airfield built for wartime use only. It is built so as to satisfy minimum military requirements.

j. Hard standing.—Surfaced or paved area used for parking of an individual airplane.

k. Landing strip.—Prepared strip of land used for landing and taking-off of aircraft. It may or may not have a runway.

l. Runway.—Paved or surfaced strip located in the center of a landing strip.

m. Shoulder.—Graded area adjacent and parallel to runway.

n. Staging field.—Intermediate landing and take-off area with a minimum of servicing, supply, and shelter, for temporary occupancy of military aircraft during movement from one airdrome to another.

o. Taxiway.—Surfaced or paved way primarily intended for circulation of aircraft on and near an airfield.

■ 80. THE AIRFIELD.—The building of a military airfield is an involved and complicated construction operation. In many respects it is like building a superhighway to support very heavy wheel loads. But there are certain differences from road-building which are extremely important, and with which the aviation engineer must be fully acquainted in order to accomplish his job.

a. Construction.—An airfield must be able to take, for the most part, a heavier load than a road. Where an average heavy load for a road is a 10-ton truck, a runway may have to support an 80-ton bomber. It is clear, therefore, that airfields must be built on firm, well-drained ground, with a strong base. [. . .]

b. Surfacing.—The surface of a runway must be smooth and even, free from pebbles or loose material that may be blown into the air and damage propellers and other parts of a plane. Since a plane lands at very high speeds, compared with vehicles, small rocks and other obstructions that would be unimportant in a road should not be allowed to remain on the runway.

Troops of the U.S. 746th Engineer Aviation Battalion resurface Lahug Airstrip on Cebu Island, Philippines, in 1945. (NARA)

c. Length of runway.—The faster and heavier a plane, the longer the runway must be. Therefore landing fields for bombers, fighters, and light aircraft are of different lengths.

d. Camouflage.—If we can reach enemy installations from our fields, they can likewise reach our fields. It is important that our fields be hard to find, and, if found, hard to see. Therefore an airfield is laid out to take advantage of natural concealment, and every attempt is made to camouflage both the airfields and the individual planes. This is an important function of the Corps of Engineers.

e. Figure 128 illustrates a typical airfield and how ground features are used to help conceal it.

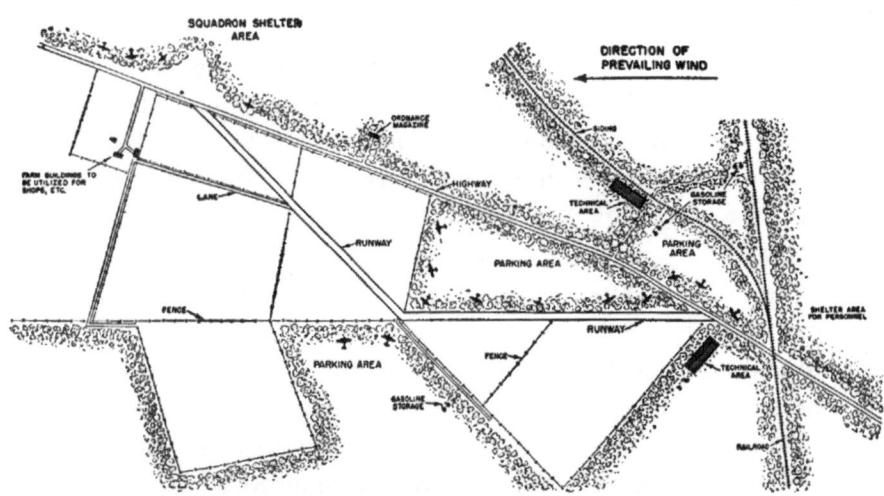

FIGURE 128.—Field airdrome. Note how runways and taxiways are laid out to fit existing road pattern, which makes it much easier to camouflage. Airplanes are widely dispersed and concealed in edge of bordering woods. One runway is always laid out in the general direction of the prevailing wind.

■ 81. STEEL RUNWAYS.—Since construction of airfields requires a great deal of time and labor, the Army has looked for ways of constructing a field quickly. For this purpose we have developed prefabricated steel landing mats of various kinds, which can be laid down quickly with a minimum of tools and equipment. It is important that every engineer know the essentials of laying down these steel runways. These runways are simple to put together, but it is up to you to know how to handle the various parts for rapid construction.

■ 82. THE AIRFIELD AND THE ENGINEER.—*a. Maintenance.*—Building the airfield is a big job, but the engineers' job does not end there. It is just as important to keep that field in a condition to be used at all times. Since the field comes under fire of various sorts, engineers must be alert and ready to fill bomb craters and to clean debris, shell fragments, and other foreign material from the runway surface. The maintenance of camouflage practice and discipline is also the engineers' job.

b. Equipment.—Aviation engineers are given much heavy machinery—bulldozers, power shovels, road graders, tractors, trucks. This material must be kept in the best of condition. These powerful machines are the engineer's tools; without them he cannot do his job. To fill a crater made by a 300-pound bomb means that 90 tons of material must be moved. With his heavy equipment, the aviation engineer can do the job in a short time; without his equipment, the job will be done too late to help the air force, too late to keep the field serviceable, too late to allow our mission to be successful. Your equipment must be ready.

■ 83. KEEP 'EM FLYING.—Like other Army engineers, the aviation engineer must do his utmost to prevent anything from hindering the forward and continued movement of our forces. Sure, it's a tough job, but engineers are tough soldiers. The construction and maintenance of an airfield is one of the stiffer challenges thrown to the engineer. We are meeting it successfully. Whether we continue to win the "battle of the airfields" depends upon how well you learn your job and upon the courage with which you carry it out. Keep 'em flying!

CHAPTER 4

Fortifications and Defenses

Although any U.S. Army infantryman required a basic understanding of how to dig a foxhole or one- or two-man fighting position, the engineers were the go-to forces for the construction of more advanced or resilient defensive structures. The February 1944 edition of FM 5-15, Corps of Engineers: Field Fortifications *brought together the accumulated wisdom not only of pre-war theory and field experience, but also knowledge gained in campaigns in North Africa, Sicily, Italy, and the Pacific, hence it would provide a solid reference guide for future Army operations in Western Europe from June 1944.*

The manual, which runs to 270 pages, dispels any notions about field fortifications being crude holes in the ground with some improvised overhead cover. Much science and engineering informed the location, layout, design, and tactical properties of even the most basic fighting position. Any field fortification had to fulfil a wide-ranging number of key requirements to be functional and effective. For example, it had to be habitable, with efficient drainage, adequate ventilation, and, if underground, adequate light. It had to provide resilient protection from enemy fire and be able to cope with predicted blast and penetration effects. For example, if the position was intended to shield the inhabitants against a 500lb general-purpose bomb, its overhead cover had to consist of either 7ft of reinforced concrete, 9½ft of stone masonry, 12ft of logs, 16ft of crushed stone, or 40ft of tamped earth. The position also had to be designed to allow the most effective use of a particular weapon system, from a single infantryman's rifle through to heavy artillery pieces, and had to facilitate smooth troop and supply movements between positions if necessary. Camouflage was another consideration—the position had to be concealed from enemy aerial and land observation.

As described below, FM 5-15 distinguishes between two types of field fortification, "hasty fortifications" and "deliberate fortifications," the first being light positions thrown down quickly to meet a sudden emergency or tactical requirement, the latter being more substantial or elaborate structures providing more durable defense or habitation. But given the almost entirely offensive mindset of the U.S. Army in World War II, most of the field fortifications described were essentially temporary in nature. Indeed, it is noteworthy that the following 1944 edition of FM 5-15 largely rid itself of previous editions' content related to long-term defensive structures, such as sophisticated trench networks; such would be a sign of operational failure in the age of maneuver warfare.

★★★

From FM 5-15, *Corps of Engineers: Field Fortifications* (1944)

CHAPTER 1
GENERAL

1. PURPOSE AND SCOPE. Troops in occupied positions increase their combat effectiveness by works of an engineering nature called field fortifications. This manual describes field fortification methods and gives details of construction of entrenchments, emplacements, and shelters. It also outlines the principles of terrain appreciation which apply to field fortifications, and explains how to combine individual field fortifications into a unified system by means of organization of the ground.

2. CLASSIFICATION AND USE OF FIELD FORTIFICATIONS.
a. Classification. There are two general classes of field fortifications.
(1) Hasty fortifications. Those initially constructed when in contact with the enemy or when contact is imminent. They consist generally of light clearing of fields of fire, foxholes for personnel, open weapon emplacements, hasty antitank and antipersonnel mine fields, barbed–wire

entanglements, strengthening of natural obstacles, observation posts, and camouflage.

(2) Deliberate fortifications. Those constructed out of contact with the enemy, or developed gradually from hasty fortifications. They include deliberate entrenchments, antitank and antipersonnel mine fields, antitank obstacles, covered weapon emplacements, barbed-wire entanglements, troop shelters which are proof against artillery fire and weather, extensive signal communication systems, gasproof inclosures of command posts and aid stations, and elaborate camouflage. [...]

CHAPTER 2
TERRAIN EVALUATION

Section I
GENERAL

3. PURPOSE. The purpose of this chapter is to describe the means of evaluating terrain. For a detailed discussion of the effect of terrain on tactical dispositions and for information on organization of the ground, see FM 100-5, 7-10, 7-15, 7-20, and 7-40.

4. DEFINITIONS. a. Terrain, from a military viewpoint, is an area of ground considered in relation to its use for military purposes.

b. Terrain evaluation is the analysis of the area of probable military operations, to determine the effect of the terrain on the lines of action open to opposing forces in the area.

5. INFLUENCE OF TERRAIN. a. The character of the area or region of military operations often has a decisive influence upon the course of operations. The more important factors to be considered in evaluating terrain include not only natural features, such as ridges, streams, bodies of water, woods, and open spaces, but also such features as roads, railways, and towns.

b. Ground forms, such as a succession of ridges and valleys, influence military operations by aiding or hampering the movement of military forces. An advance parallel to the ridges and valleys is mechanically easier than movement across successive ridges.

c. The salient feature of a commander's plan of action are usually determined so as to take full advantage of favorable terrain features.

6. TERRAIN FACTORS. Regardless of the type of terrain and the tactical situation, terrain always can be evaluated in terms of the following five factors: observation, fields of fire, concealment and cover, obstacles, and communications.

a. Observation. Observation of the ground on which a fight is taking place is essential in order to bring effective fire to bear upon the enemy. Observation also aids in increasing the effectiveness of fire directed on an enemy stopped by obstacles. The value of cover and concealment is based on denial of observation to the enemy. Observation also affords information as to what both enemy and friendly troops are doing, making it possible for the commander to control more effectively the operations of his troops.

b. Fields of fire. Fields of fire are essential to the defense. An ideal field of fire for infantry is an open stretch of ground in which the enemy can be seen and in which he has no protection from fire as far as the limits of effective range of the infantry weapons. Fields of fire can be improved by cutting or burning weeds, grass, and crops; clearing brush and trees; demolishing buildings; and cutting lanes through woods. However, concealment must be carefully considered in all such work. The time and labor available for this type of improvement should be considered in evaluating the terrain.

c. Concealment and cover. Concealment from view, both from the air and ground, will usually protect military personnel and installations only as long as the enemy is unaware of their location. Unconcealed installations and troops invite destruction. Cover includes protection from fire, either that provided by the terrain, or that provided by other natural or artificial means.

d. Obstacles. Obstacles are obstructions to the movement of military forces. Some of the common natural obstacles of military value are mountains, rivers, streams, bodies of water, marshes, gullies, steep inclines, and heavily wooded terrain. Proper evaluation of natural obstacles permits the most effective use of artificial obstacles.

(1) Mountains parallel to the direction of advance of a force limit or prohibit lateral movement and protect the flanks; perpendicular to the advance, they are an obstacle to the attacker and an aid to the defender.

(2) Rivers are similar to mountains in their effect on forces moving parallel and perpendicular to them. Rivers flowing parallel to the advance may be used as routes of supply.

(3) Marshes frequently provide more delay to an advance than bodies of water, because generally it is more difficult to build causeways than bridges. Mechanized vehicles can be restricted in movement by dense woods, marshes, steep inclines, gullies, stumps, large rocks, and bodies of water.

e. Communications. Communications consist of roads, railroads, waterways, airways, and their facilities. They are important to both offense and defense for moving troops and supplies. In most situations, especially in the operations of large bodies of troops, the means of communication are of vital importance. The existing ones generally must be studied thoroughly and utilized to the maximum before new ones are constructed.

7. OBJECTIVES. Terrain objectives, normally, are clearly defined features, the capture of which will insure the defeat of a hostile force, or from which the operation can be continued or the success exploited. Terrain objectives, in the attack by ground forces, usually are located in, or in rear of, the hostile artillery area. One may be a terrain feature affording command observation, another a critical point in the hostile command system or on essential supply routes, and another an obstacle to armored forces. In some situations the objective is clearly indicated by the mission; in others it is deduced from the situation.

8. MAPS AND RECONNAISSANCE. Maps are the basis for terrain studies but should be checked by air reconnaissance, air photographs, and ground reconnaissance. Works of man, especially routes

Many black soldiers – such as those of the color guard seen here – served in the engineers during World War II, although like all black servicemen they suffered from segregation and prejudice. (LOC)

of communication, are changing constantly; and even natural ground forms may change.

9. CORRIDORS. Features such as ridges, streams, woods, roads, and towns divide all terrain into more or less separate areas. Such an area frequently consists of a valley lying between two ridges or an open space between two wooded areas. The limiting features prevent direct fire or ground observation into the area; they may be high or low, continuous or discontinuous. When the longer axis of such an area extends in the direction of movement of a force, or leads toward or into a position, the area is called a corridor.

[. . .]

CHAPTER 3
GENERAL FORTIFICATION TECHNIQUE

SECTION II
GENERAL TECHNIQUE

20. CLEARING FIELDS OF FIRE. Suitable fields of fire are required in front of each entrenchment or emplacement. In clearing them the following principles must be observed:

a. Do not disclose position by excessive or careless clearing.

b. In areas organized for close defense, start clearing near main line of resistance and work forward at least 100 yards.

c. In all cases leave a thin natural screen to hide defense positions.

d. In sparsely wooded areas, remove the lower branches of scattered, large trees. Occasionally it is desirable to remove entire trees which might be used as reference points for enemy fire.

e. In heavy woods, complete clearing of the field of fire is neither possible nor desirable. Restrict work to thinning undergrowth and removing lower branches of large trees. In addition, clear narrow lanes for fire of automatic weapons.

f. Remove or thin thick brush. It is never a suitable obstacle and obstructs the field of fire.

g. Demolish other obstructions to fire, such as buildings and walls, only when resulting debris provides less enemy protection.

h. Mow grain crops and hay fields or, if ripe and dry, burn them if it will not disclose the position. Usually this is practicable only for a deliberate position organized prior to contact with the enemy.

i. Drag away cut brush to points where it will not furnish concealment to the enemy nor disclose the position.

j. Before clearing the fields of fire make a careful estimate as to how much clearing can be done in the time available. This estimate often determines the nature and extent of the clearing to be undertaken, since a field of fire only partially cleared may afford the enemy better concealment and cover than the area in its natural state. Estimates may be based on table I, which makes no allowance for the removal of debris.

Additional allowance must be made for this, depending upon tile amount of debris, length of haul, and equipment available.

Table I. Man-hours required to clear 100 square yards

Description of clearing	Tools used	Man-hours required
Medium clearing—clearing under-growth and some trees not exceeding 12 inches in diameter.	Saws, axes	3½
Light clearing—clearing small brush only.	Axes	1½

21. CAMOUFLAGE. Concealment is of prime importance in locating defensive works. Before any excavation is started, all turf, sod, leaves, or forest humus is removed carefully from both the area to be excavated and that on which spoil is to be piled. This material is set aside and replaced over the spoil when the work is completed. To prevent discovery of the work during excavation, camouflage nets are suspended from stakes or trees before excavation is started. The workers confine their activities to the area beneath the camouflage net. The net is suspended high enough above the ground to permit excavation without snagging equipment or entrenching tools on it. After the excavation has been completed and the spoil covered with sod or other natural camouflage material, the net is lowered close to the ground so that it is inconspicuous from ground observation. Nets are kept in position when the weapon is not being fired. Arrangements are made to withdraw or lift the net during action.

22. EXCAVATION. Excavation is usually by pick and shovel. The nature of the soil, tools available, condition and experience of men, presence of the enemy, amount of light to work by, size of excavation, and weather conditions, all affect the rate of excavation. Because of the large number of variables involved, precise data cannot be given. As a rough guide, it may be stated that in medium soil, using standard size tools, a man in good condition can excavate between 20 and 30 cubic feet per hour. Table II gives estimates for excavation of man-hours required to build various types of infantry weapons emplacements.

TABLE II. *Excavation and camouflage data for infantry weapons emplacements*

Weapon	Type of emplacement*	Area to be camouflaged (feet)	Excavation (cubic feet)	Man-hours required to construct in medium soil
Rifle	Foxhole	10 x 10	37	1½
Automatic rifle	do	10 x 10	37	1½
Rocket launcher	Pit-foxhole	10 x 10	{ 125 / 287	1 / 4½
	Pit	5 x 5	50	3
Machine gun, light, cal. .30.	Horseshoe	15 x 15	123	7
	2-foxhole	12 x 12	74	3
Machine gun, heavy, cal. .30.	Horseshoe	15 x 18	140	8
	3-foxhole	15 x 15	111	5
60-mm. mortar	Pit	14 x 14	70	4
81-mm. mortar	do	16 x 16	108	6
37-mm. AT gun	Circular	21 x 21	110	5
	Fan	29 x 29	195	10
	Rectangular	27 x 36	550	28
57-mm. AT gun	Fan	24 x 39	410	21
105-mm. howitzer M3	Circular	29 x 29	430	22

*See chapter 4 for details of emplacements. ¹ Without foxholes. ² With foxholes.

23. DISPOSAL OF SPOIL. Excavated soil is much lighter in color and tone than surface soil and must be hidden carefully lest its presence disclose the fortification. There are several ways to dispose of spoil.

a. It may be used to form a parapet if the topsoil is carefully saved and used to cover the parapet. Turf, sod, leaves, or other litter from under nearby bushes or trees are used to make the parapet resemble its surroundings.

b. It may be removed and carefully hidden under trees or bushes or in ravines. Care must be taken to avoid revealing tracks.

c. It may be collected and used, partly camouflaged, to form parapets for dummy positions.

24. DRAINAGE. Lack of proper drainage increases the maintenance work and the hardships of the troops occupying the fortifications. It must be provided for in the lay-out and construction of all works.

a. Proper location. Proper location limits but does not eliminate the drainage problem. If possible, low points and drainage lines are avoided, and trenches are located on slopes. A slope of 1 percent causes all water to run to the lowest part of trench, from which it can be easily drained. Slopes exceeding 2½ percent cause erosion.

b. Surface and rain water. Surface and rain water can be largely excluded by deflecting it through the use of small ridges into ditches passed around the fortification, or over it by means of flumes.

c. Subsurface water. Surface water may be removed by the use of sumps or of drainage ditches run to natural drainage lines from low points in entrenchments or emplacements. Sumps are located at low points and emptied by bailing, siphoning, or pumping. They should be a minimum of 1½ feet square and 1 foot deep.

25. REVETMENTS. A revetment is a retaining wall, or facing, for maintaining an earth slope at an angle steeper than its natural angle of repose. In all but the hardest ground, when the position is to be occupied for more than a few days, some measures must be taken to prevent crumbling of walls. Decreasing the slope for this purpose also decreases protection afforded by the emplacement and makes concealment more difficult. Revetments require considerable labor and material, but reduce maintenance and insure stability. Earth walls in entrenchments and emplacements not only are subject to normal erosion processes and wear and tear of constant occupation, but also must withstand heavy earth shock caused by explosion of bombs and artillery shells. There are two kinds of revetment, the retaining-wall type and the surface or facing type.

a. Retaining-wall type. This type is self-supporting and acts by its weight. Dimensions of the excavation must be increased to allow space for this type of revetment.

(1) Sandbag revetment. These are particularly useful for emergency work, for repairs, and on the interior slopes of earth parapets.

(a) The standard sandbag measures 14 by 26½ inches when empty, and has a string attached 3 inches from the top. When three-fourths full, the bag weighs from 40 to 75 pounds, depending upon nature and moisture content of the filler. The average filled sandbag weighs about 665 pounds, and occupies a space 4¾ by 10 by 19 inches. The following data are useful in estimating the number of sandbags required for reveting purposes:

1. If a single row of stretchers is used, as occasionally is done for small revetments, about 160 sandbags are required for each 100 square feet of surface to be reveted.
2. If alternate headers and stretchers are used, as is proper, about 320 sandbags are required for each 100 square feet of surface to be reveted.
3. If sandbags are used for fills, parapets, or breastworks, about 195 are required for each 100 cubic feet of fill.

(b) Ordinary sandbags should be used for temporary reveting only. Where bags are to be in place for a month or longer under average conditions of moisture, they must be rotproofed or filled with soil partially stabilized with cement or bitumen. The latter method usually is simpler in the field.

(c) Sandbags are laid as follows:

1. Fill bags uniformly about three-fourths full.
2. Tuck in bottom corners of bag after filling.
3. Build walls with slope 3 on 1 to 4 on 1.
4. Place bags perpendicular to slope.
5. Place bottom row headers.
6. Alternate intermediate rows as headers and stretchers.
7. Complete with a top row of headers.
8. Place side seams and choked ends on the inside.

9. Break joints and beat bags into place and into rectangular shape with back of shovel, or tamp with feet.

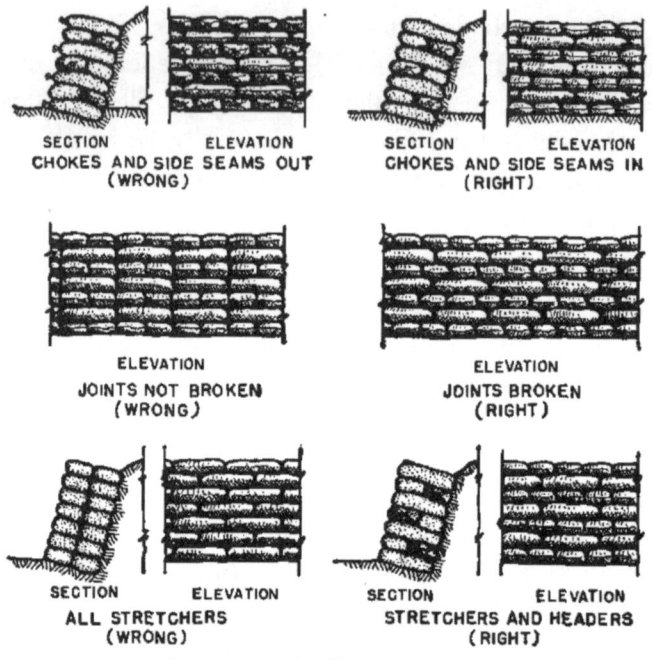

Figure 11. Sandbag revetment.

(2) Sod revetment. Thick sod makes durable revetments. Sods are cut about 18 by 9 inches, laid grass down, except for top layer, and pinned together with wooden pegs. The procedure given for sandbag revetment applies.

b. Surface or facing revetment. Surface or facing revetment must be supported, and serves mainly to protect the reveted surface from effects of weather and damage due to occupation. When strongly constructed, it retains loose material. Its top should be about 8 inches below the ground level to prevent its being snapped or damaged if tanks cross the reveted wall.

(1) Issue material. Issue material, such as burlap and chicken wire, wire mesh, expanded metal (XPM), or corrugated iron, for this type of revetment may be obtained in limited quantities at engineer dumps. These materials are held in place against the surface to be reveted either by wooden pickets (at least 3 inches in diameter) or by issue steel pickets. Pickets are driven into the floor and held at the top by holdfasts. In installing this type of reveting, the following operations are necessary:

(a) Cut grooves for pickets into wall to be reveted. Space 1½ to 6 feet apart, depending upon reveting material to be used.

(b) Prepare holdfast in front of each groove. Hold fast anchor picket should be from 8 to 10 feet from wall.

(c) Place two end pickets loosely. Stretch material between them and hold taut while end pickets are tightened.

(d) Drive all pickets at least 1½ feet into floor and fasten tops to anchor pickets with two turns of No. 10 wire. Draw pickets tight by racking. Pickets draw material tight against surface to be reveted.

(2) **Natural material.** Since issue material often is difficult to obtain in the field, most reveting is done with natural material such as brush and cut timber obtained at the site.

(a) *A brushwood hurdle* is a woven revetment unit usually 6 feet long and of required height. Brushwood less than 1 inch in diameter at butt is woven on a framework of sharpened pickets driven into the ground at 18-inch intervals. When finished, hurdle is driven into floor and held in place by holdfasts.

(b) *Continuous brush revetment* is constructed by driving 3-inch pickets at 1-pace intervals about 4 inches from face of surface to be reveted. Space behind the pickets then is packed with small, straight brush laid horizontally. Pickets are drawn tight by means of holdfasts.

(c) *Cut-timber revetment* is the principal natural means of reveting foxholes and emplacements. It is similar to continuous brush revetment, except that a horizontal layer of small timbers, cut to length of wall to be reveted, is used in place of brush. Pickets are held in place by holdfasts or struts. When available, dimension lumber may be used in a similar manner.

26. BREASTWORKS AND PARAPETS. Breastworks and parapets are built for protection when soil conditions or subsurface water prevent excavation. They also are used with dug-in emplacements to save extra digging. They are built at least 3 feet thick at the top to protect against caliber .30 bullets and against shell fragments. Also, they should be free of loose rocks and pieces of wood.

27. PROTECTION AGAINST TANKS. So far as practicable, entrenchments and emplacements are built to provide protection against tanks. Rocks or other pieces of hard material should not be left on or near the surface of the ground within 3 or 4 feet of the lip of the entrenchment or emplacement. The pressure on such a rock caused by a tank collapses earth walls which otherwise would be able to withstand the passage.

CHAPTER 4
ENTRENCHMENTS AND EMPLACEMENTS

Section II
INFANTRY ENTRENCHMENTS FOR HASTY FORTIFICATIONS

30. GENERAL. Entrenchments are located to cover a selected area with fire and, at the same time, provide concealment from aerial and ground observation and protection from enemy fire. These requirements are met when the troops are located in one- or two-man foxholes as described in this manual. To confuse the enemy, judicious use must be made of decoys or dummy positions.

31. FOXHOLES. a. General. Foxholes are entrenchments normally dug for individual protection when contact with the enemy is imminent or in progress. They provide excellent protection against small-arms fire, artillery shell fragments, airplane fire or bombing, and the crushing action of tanks. The one- and two-man foxholes are basic types, the choice of type resting with the squad leader if not prescribed by higher authority.

The two-man foxhole is used when men must work in pairs or when, for psychological reasons, battlefield comradeship is desirable.

b. Use. For units within the battle position, foxholes are sited with the longer side generally parallel to the front, but they are distributed around weapon emplacements to provide for all-around defense. Troops occupy their foxholes only when an attack is imminent or in progress. In some situations, where the need for rest is paramount, commanders may permit soldiers to stop excavation before full depth has been reached.

32. ONE-MAN FOXHOLE. a. Dimensions.

(1) The size and shape of the foxhole are affected by the following:

(a) It is as small as practicable, to present the minimum target to enemy fire.

(b) It is wide enough to accommodate the shoulders of a man sitting on the firestep.

(c) It is long enough to permit the use of large-size entrenching tools.

(d) It is at least 4 feet deep to the firestep, from which the standing occupant should be able to fire.

(e) A sump is dug in one end for bailing out water and for the feet of the seated occupant.

(2) The foregoing considerations result in the dimensions shown in figure 17. The soldier should memorize these simple dimensions: 2 feet wide, 3½ feet long, 4 to 5 feet deep depending upon the height of the man, and additional depth at one end for the sump.

b. Details of construction. In most types of soil the foxhole gives positive protection against the crushing action of tanks, provided the soldier crouches at least 2 feet below the ground surface. In very sandy or very soft soils it may be necessary to revet the sides to prevent caving in. The spoil is piled around the hole as a parapet, 3 feet thick and approximately ½ foot high, leaving a berm or shelf wide enough for the soldier to rest his elbows upon while firing. If turf or topsoil is to be used to camouflage this parapet, the soldier first removes the topsoil from an area 10 feet square and sets it aside until the foxhole is completed.

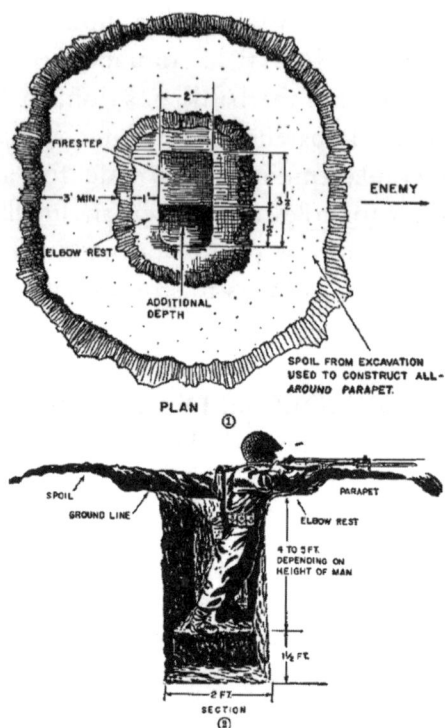

Figure 17. One-man foxhole. (Camouflage omitted.)

c. Foxhole with camouflage cover. It may be practicable for the soldier to remove the spoil to an inconspicuous place and to improvise a camouflage cover for his foxhole. This technique is especially effective against a mechanized attack supported by foot soldiers. Riflemen remain concealed until the tanks have overrun the position; then they rise up and combat the enemy foot soldiers following the tanks.

33. TWO-MAN FOXHOLE. The two-man foxhole consists essentially of two adjacent one-man foxholes. Since it is longer than the one-man type, the two-man foxhole offers somewhat less protection against tanks crossing along the long axis, as well as against airplane strafing and bombing and artillery shell fragments.

[. . .]

SECTION **III**

INFANTRY WEAPON EMPLACEMENTS

[. . .]

39. CALIBER .30 MACHINE GUN (LIGHT). There are two types of emplacements for this gun: the horseshoe type and the two-foxhole type.

a. Horseshoe type. (1) The gun is placed in firing position ready for immediate action. Lying down, if exposed to fire, the crew first excavate about ½ foot beneath the gun and then a similar depth for themselves, thus making an open shallow pit. The spoil is piled around in a parapet.

(2) The emplacement is completed by digging out a horseshoe-shaped trench, about 2 feet wide, along the rear and sides of the pit, leaving a chest-high shelf to the center and front to serve as a gun platform. The spoil is piled around the emplacement to form a parapet at least 3 feet thick and low enough to permit all-around fire.

(3) This emplacement furnishes protection against small-arms fire and shell or bomb fragments. In firm soil, this emplacement offers protection against the crushing action of tanks. In loose soil, logs about 8 inches in diameter, placed across front, rear, and sides of the emplacement and embedded flush with the top of the ground, help to make the emplacement resistant to the crushing action of tanks. When tanks appear about to overrun the position, the gunners pull the weapon to the bottom of the trench at the rear of the emplacement and then crouch down to either side.

b. Two-foxhole type. This emplacement consists of two one-man foxholes close to the gun position. To lay it out, a short mark is scratched on the ground in the principal direction of fire. On the right of this mark a foxhole is dug for the gunner. On the left of the mark and 2 feet to the front, another foxhole is dug for the assistant gunner. The spoil is piled all around the position to form a parapet, care being taken to pile it so as to permit all-around fire of the weapon. In firm soil, the two-foxhole type provides protection for the crew and the weapon against the crushing action of tanks. When tanks appear about to overrun the position, the gun is removed from the tripod and taken into one foxhole, the tripod into the other. The gunner and assistant gunner crouch in the holes.

c. Choice of type. As a firing position, the two-foxhole type is a little less flexible than the horseshoe type, but it is easier to construct and more nearly tankproof than the horseshoe type. Therefore, the two-foxhole type generally is preferred.

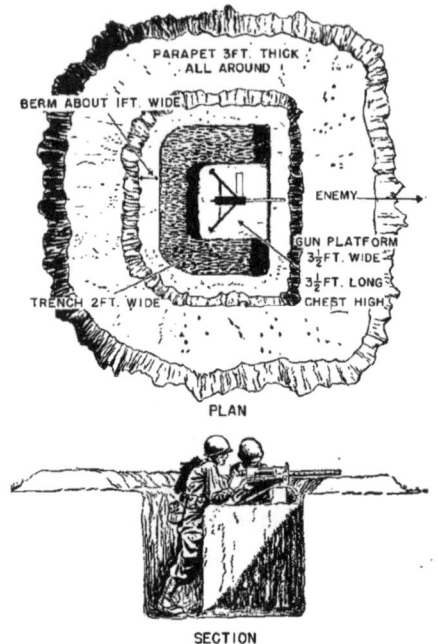

Figure 25. Horseshoe type emplacement for caliber .30 machine gun (light). (Camouflage omitted.)

[. . .]

41. 60-MM MORTAR EMPLACEMENT. a. Open type. This consists of a rectangular pit large enough to accommodate the mortar, the gunner, and the assistant gunner. The emplacement is kept to the minimum size to afford protection against airplane fire and bombing and against artillery shells, but it allows room for firing the mortar and storing necessary ammunition. The front edge is sloped so that the aiming stake, about 10 yards to the front, is visible through the sight and so the weapon's fire will be clear. The spoil from the excavation is piled all around the pit to form a low parapet. Foxholes for members of the mortar squad not required at the gun are prepared not far from the emplacement. Additional ammunition is placed in nearby shelters.

b. Two-foxhole type. [. . .] The two-foxhole type of emplacement is preferred when the mortar is in defilade.

42. 81-MM MORTAR EMPLACEMENT. Except for somewhat larger dimensions, this emplacement is the same as the open type described above for the 60-mm mortar. A revetted ammunition niche may be built into the side of the pit.

43. ROCKET LAUNCHER. There are two types of emplacement for this weapon, the pit-foxhole type and the pit type.

a. Pit-foxhole type. This emplacement is a circular pit, 3 feet in diameter and about 3½ feet deep, large enough for two men. It permits the assistant rocketeer to turn with the traversing weapon, so that he is never behind it when it is fired. The emplacement is shallow enough to permit the rear end of the rocket launcher at maximum elevation to be clear of the parapet, thus insuring that the hot back-blast from the rockets is not deflected to the occupants. This emplacement is not tankproof. Therefore foxholes for the crew are dug nearby. As the antitank mission of this weapon requires that it be kept in action against hostile tanks until the last possible moment, these foxholes will be occupied only when a tank is about to overrun the emplacement.

b. Pit type. In firm soil the diameter of the circular pit can be increased to 4 feet and an additional circular pit 2 feet deep and 2 feet in diameter excavated in the center. This leaves a circular fire step 1 foot wide and about 3½ feet below the surface. When tanks appear about to overrun the position, the rocketeer and assistant rocketeer crouch down into the lower pit. When the tanks have passed, the rocket launcher quickly is returned to action.

44. 37-MM ANTITANK GUN EMPLACEMENT. For this weapon there are three standard types of emplacement, each adapted to a special situation. For flat terrain the circular emplacement is preferred, since it permits all-around fire. For sloping terrain the rectangular-pit emplacement with ramp is preferred, since it gives partial defilade and protection against ground observation. When the direction of enemy approach can be foreseen definitely, as in covering a road block in a defile, the fan type emplacement is preferred, since the gun can be fired instantly without being moved from the cover position.

a. Circular type emplacement. This emplacement consists of a circular pit, 11 feet in diameter and 1 foot deep, measured from ground level, with an all-around parapet 4 feet wide and approximately 9 inches high. The banks should be sloped so that the gun can be moved into and out of the emplacement. Pits for the wheels and a slit in the parapet for the gun barrel allow the gun to be lowered below the parapet when not firing, at which time the crew takes cover in nearby foxholes. An ammunition pit may be dug in the center of the emplacement or into the parapet. Additional ammunition is placed in nearby shelters. The gun is maneuvered quickly to fire in any direction by elevating the barrel, lifting the trails, backing the gun to the center of the emplacement, turning it in the desired direction, and pushing it forward against the bank.

b. Rectangular pit type emplacement. This emplacement consists of a rectangular pit 10½ feet wide, by 14 feet long, and 3½ feet deep. To get the gun into and out of the pit, a ramp is dug either straight out of the pit or at an angle thereto, depending upon the sector of fire assigned to the gun. If the ramp is turned at an angle to the pit, the elbow is curved so that the gun can be moved around the corner. The gun is backed into the rectangular pit, when not in firing position. The spoil is piled alongside the excavation to form a parapet. Several foxholes may be dug inside the emplacement without interfering with the movement of the weapon. Additional foxholes are dug nearby for the remaining members of the gun crew.

c. Fan type emplacement. This type of emplacement permits little traverse but great speed in going into action. The ramp is to the rear of the emplacement.

[…]

CHAPTER 5
SHELTERS

SECTION I
GENERAL

66. SCOPE. This chapter covers various types of shelters employed in combat operations. Included is a discussion of tactical and technical requirements of shelters, methods of construction, and type designs. [...]

67. CLASSIFICATION. a. Based on degree of protection. Protected shelters are classified according to the degree of protection they afford.

(1) Blastproof and splinterproof shelters. These shelters protect against rifle and machine-gun fire, grenades, and light mortars; splinters of most high-explosive shells and bombs; and blasts of 100-pound bombs exploding not closer than 50 feet. They do not protect against direct hits by bombs or artillery projectiles.

(2) Light shellproof shelters. Light shellproof shelters protect against direct hits by 105-mm shells and fragmentation bombs.

(3) Shellproof shelters. Shellproof shelters protect against continuous bombardment of shells up to 8 inches and direct hits by bombs up to 200 pounds, as well as against blasts of bombs up to 500 pounds exploding not nearer than 25 feet.

b. Based on method of construction. According to the method of construction, which depends on the character of the ground, materials available, and protection required, shelters may be further classified as follows:

(1) Surface. Surface shelters have maximum observation and exit facilities and require a minimum of labor; on the other hand, they are relatively conspicuous, require considerable cover material, and provide the least protection of the shelters mentioned here. They are seldom used for the protection of personnel in advanced positions unless they can be concealed in woods, on steep reverse slopes, or among buildings; or unless the underground water level is so close to the surface that the cut-and-cover type of shelter cannot be used. Shelters consisting of almost any type of small, improvised shed, covered with a layer of earth, may be used for the protection of ammunition and stores. These shelters should be of small capacity, well dispersed, and carefully concealed.

(2) Cut-and-cover. (a) This type, intermediate between the surface and the cave shelter, is constructed in an open excavation which is backfilled around and over the structure to the level of the original surface, or somewhat above. The resisting power of the overhead cover is increased by layers of concrete, steel beams, broken stones, or other materials with high resistance to penetration.

(b) The cut-and-cover shelter is better adapted than the cave shelter for use as a dressing station. It is more quickly constructed, more easily cleaned, better ventilated, and offers easier means of admission and evacuation of casualties. However, it usually requires much larger quantities of materials to provide the same protection, does not resist intensive shelling as well, and is harder to conceal.

(c) When the use of cave shelters is impractical because of surface or underground water, hardness of underlying rock, or the rapidity of exit required, cut-and-cover shelters sometimes may be used. They are also suitable in wooded areas or in buildings, where concealment is easy and ample material is available, and in situations requiring immediate shelter.

(d) Unless they are constructed partially or wholly of concrete, cut-and-cover shelters do not offer much protection against shells heavier than the 6-inch type.

(3) Cave. (a) Cave shelters are constructed entirely below the surface by mining methods, and have a cover of undisturbed earth. They are the least conspicuous of all types, afford effective protection before completion, and require the minimum material. Their disadvantages are: limited observation, congested living conditions, small exits, difficult drainage and ventilation, and time required to build.

(b) It is difficult to increase overhead protection of these shelters after completion, since protection depends upon depth at which the chamber is built. For this reason, in determining the depth, it is important not to underestimate the protection needed. On the other hand, it is equally important not to overestimate it because of the extra time, labor, and material involved.

68. TACTICAL CONSIDERATIONS AND REQUIREMENTS.

a. Purpose. The primary purpose of a protected shelter is to permit troops or important installations to remain in comparative safety at or near their combat positions during hostile bombardment.

b. Terrain. Reverse slope positions, since they are difficult for artillery to hit and usually are easily drained, make excellent locations for shelters. Wooded areas and buildings, which provide materials and facilitate concealment, are also desirable.

c. Location. It is of the utmost importance near the front to have the shelters near where the troops who occupy them are needed. This rule is relatively less important toward the rear. Facilities for cover and concealment also influence the location of shelters. Every advantage which the tactical situation permits should be taken of natural shelter.

d. Ease of exit. (1) The occupants of a shelter must be able to get out rapidly. This is particularly important in shelters located near the front, where troops must be able to get out and occupy their fighting positions in the small time available between the enemy bombardment and the infantry assault.

(2) Exit is made easier by designing shelters with small capacity, minimum depth below ground, and unrestricted entrances.

(3) Large shelters are provided with at least two entrances, and preferably with a third for emergency use. Small shelters are provided with exits as necessary. Extra exits should emerge at a different place from the main exit and where practicable at points well concealed or camouflaged. Entrances should be spaced a minimum of 40 feet apart to avoid the danger of one shell or bomb burst blocking two entrances. Large systems of cave shelters should provide one entrance for every 25 men.

e. Concealment. It is important that the location and number of shelters be concealed from enemy air and ground observation. Changes in the appearance of the terrain must be avoided. Materials and spoil must be concealed or camouflaged as the shelters are built. Strict camouflage discipline must be enforced among working parties. Surface shelters may be hidden by terrain features such as woods or buildings. Cut-and-cover shelters should be kept low.

f. Observation. If practicable, shelters should be located to afford necessary observation, and should be provided with means of observation, for example, loopholes in a surface shelter, or a camouflaged periscope in the roof of a cut-and-cover shelter.

g. Application of types. Because of the time element and the construction difficulties in mobile warfare, the blastproof and splinterproof shelter is the type usually used. The necessity for shelter becomes greater as stabilization develops and details of the position become known to the enemy. In the rear parts of the defended area, larger and deeper shelters are both permissible and economical. These usually accommodate one or two squads or a platoon. They may be developed from the emergency shelters initially constructed. Shellproof and cave shelters are used only in stabilized situations.

h. Requirements for shelter in advanced positions. (1) Shelters in the advanced lines should be—

(a) Well distributed, placing troops close to their combat positions.

(b) Constructed without going to great depths, to provide for ease of exit.

(c) Provided with direct and easy exits, even at some sacrifice of cover.

(d) Of small capacity (from two to twelve men).

(e) Of a type that can be constructed rapidly.

(f) Concealed as thoroughly as possible.

(2) These requirements usually limit the type to the blastproof and splinterproof shelter.

i. Requirements for shelters in rear positions. Shelters in rear positions may be larger and deeper than those at the front. Occupants usually have more time to emerge after warning of attack. Also in these areas shelters can be given maximum overhead cover to withstand bombardment of light bombs and heavy shells, giving occupying troops the necessary rest and a feeling of security. If underground water conditions permit, the shelters are built entirely below the ground. They are carefully hidden from enemy aerial observation.

69. TECHNICAL CONSIDERATIONS AND REQUIREMENTS. These include—

a. Subsurface conditions. Subsurface conditions such as extent and character of underlying rock, position and thickness of impervious and water-bearing strata, and amount of water to be controlled.

b. Facilities available. Facilities available, including time, personnel, tools, material, and transportation.

c. Drainage. Drainage of deep shelters sometimes becomes a complex problem. It includes exclusion of surface and rain water from the entrance, exclusion of seepage from the interior, and removal of water that has collected in the interior.

(1) Surface and rain water. Surface and rain water must be excluded from all shelter entrances. If shelter is enforced from a trench and drainage is sluggish, two sumps may be dug in the bottom of the trench, at least 6 feet from the sides of the entrance, and strongly revetted. The bottom of the trench in front of the entrance must then be graded to the sumps so that the highest point is in front of the entrance. At times it is possible to dig a sump in front of the entrance and grade the trench so that only a limited portion drains into it. Direct rainfall into entrances is prevented either by the design of the entrance or by the construction of some form of weatherproof shelter above it. Baffle boards placed at the entrance floor are useful to keep out surface water.

(2) Seepage. Protection against water seeping into shelters is important. In a surface or cut-and-cover shelter this is accomplished by placing tarpaper between the sheeting and the cover. In cave shelters a strip of corrugated iron may be placed on the cap of the frame. Sheeting is then driven over the iron. Space between caps is filled with an additional piece of corrugated iron supported by struts. Seepage is thus carried to the sides of the chamber, where it collects in a ditch leading to a sump.

(3) Removal of water from chambers and galleries. Galleries should be driven on a 1-percent (or 1 foot to 100 feet) grade longitudinally, and all slopes should fall toward a point or points where the water can be disposed of. If the shelter has a level entrance, regulate slopes so that water will run to the mouth. The gallery floor should slope laterally in a 1-percent grade and a ditch should be dug along one side. In a shelter entered by incline or shaft, a sump must be formed at the bottom from which water can be removed by pumping, siphoning, or bailing.

d. Ventilation. (1) Importance. Ventilation is a particularly important factor in cave shelters. It includes the following problems:

(a) Providing sufficient circulation of fresh air in the incline, shafts, galleries, and chamber.

(b) Gasproofing, or exclusion of gas from all parts of the shelter.

(c) Providing pure air by means of air purifiers (collective protectors) when the entrances and ventilation shafts are closed during a prolonged gas attack.

(2) Circulation of fresh air. (a) In surface and cut-and-cover shelters sufficient fresh air usually is obtained by keeping entrances open.

(b) In cave shelters ventilating shafts usually are necessary in addition to entrances. They are small vertical shafts, which may be bored from within after the completion of the shelter. A stovepipe through a shaft materially assists circulation of the air. In very large and elaborate systems of shelters a draft may be created by fans.

(c) A gallery should not be driven more than 60 feet without artificial ventilation. A gallery with a single opening is ventilated by forcing fresh air to the working end through a duct of wood, metal, or canvas. A pressure blower worked by hand or power is an essential item of mining equipment. For excavations of moderate extent a portable forge forms an expedient ventilating device. Drill holes through the roofs of galleries promote ventilation. In a system of galleries having two or more outlets, air may be forced out from one outlet and drawn in through another. Screens or doors may be arranged to guide the distribution of fresh air. Vacuum operation is never as satisfactory as a pressure system.

(3) Unventilated shelters. Shelters not provided with collective protectors should be used only by personnel who are to remain inactive during occupancy. Since an inactive man requires about 1 cubic foot of air per minute, the capacity of unventilated shelters is limited in part by the difficulty of providing unusually large shelters. Initial air-space requirements for shelters for not over 12 men are 150 cubic feet per man.

[. . .]

71. OVERHEAD COVER. Figure 72 shows a typical overhead cover for protection against penetration and explosion of projectiles.

a. Bursting layer. The bursting layer covers the entire top of the shelter and extends beyond the top a distance equal to the depth of the shelter floor below the ground. The burster layer is made of standard bursters, standard reinforced concrete beams, rubble masonry, or poured concrete.

b. Distributing layers. The distributing layers tend to distribute the effects of explosion. The lowest distributing layer also bears the weight of the overhead cover and transmits it to the berms of natural soil. The minimum length resting on the berm is equal to the thickness of the bottom cushion layer, plus 1 foot. Logs wired together, steel I-beams, rails, or concrete beams set on edge, are used in distributing layers.

c. Cushion layers. Cushion layers between bursting layers and distributing layers are made of sand, gravel, tamped earth, crushed rock, or brick rubble. Preferably the top layer is of a granular material such as gravel or crushed rock, and the other layers are of tamped earth.

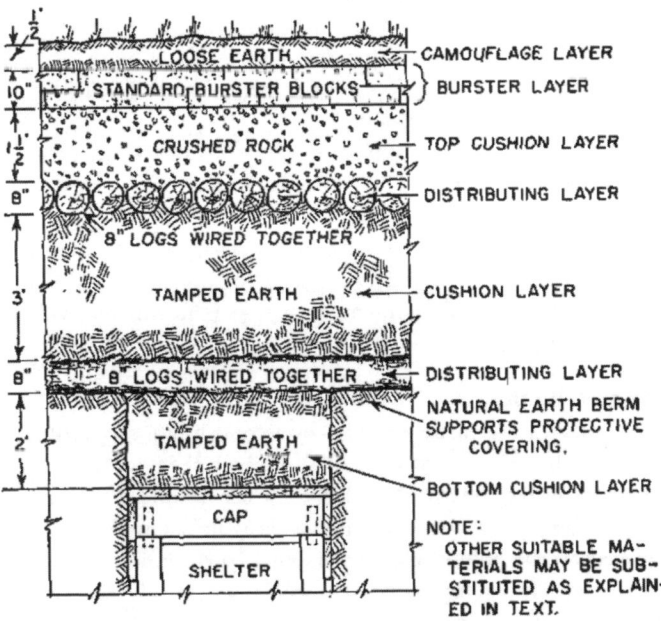

Figure 72. Typical overhead cover for cut-and-cover shelter.

German Engineer Assault Tactics

As a general rule, German engineers were offensive in orientation, and thus were highly trained in the tactics and technologies used to destroy the very strongpoints U.S. engineers might construct. Although the Western allies developed little in the way of large-scale permanent fortifications, the combat frontlines would have been extensively etched with trenchworks and punctuated by protected structures, such as pillboxes. The following text is from "Engineer Assault Tactics (German)" in a 1943 edition of the U.S. Intelligence Bulletin, explaining what the U.S. authorities knew about German principles of attack on defensive positions.

1. INTRODUCTION

This section deals with German engineer assault tactics developed since the battle of Crete. Since the information has been obtained from German prisoners of war (engineers) captured in Tunisia, it should be accepted with the reservations customary under such circumstances.

2. ATTACKS ON PILLBOXES

a. Composition of Detachment

An engineer assault detachment (*Pioniersturmzug*), whose principal task is the assaulting of pillboxes, may be composed of any or all of the following:

(1) From two to six men with pole charges or tube charges. (These tubes are said to be about 2 yards long.)

(2) From one to three flame-thrower teams of two men each. There is also said to be a third man, who accompanies them and serves as an alternate, if needed.

(3) From one to four men with hollow charges and explosives.

(4) Light machine-gun covering detachments.

The engineer assault detachment in action is normally divided into two sections (*Gruppen*).

b. Assault Tactics

The assault is normally preceded by a concentration of artillery fire. One purpose of this fire is to make craters in which the advancing engineers can take cover. When the assault detachment reaches the wire surrounding the enemy pillbox, Very signals are fired, calling for all available artillery fire to be placed on the pillbox and its immediate surroundings.

It is reported that, at this point, a smoke screen is laid by two men of the detachment, using smoke grenades (similar to stick grenades), smoke candles, or smoke canisters. Also, there are reports that smoke screens are put down as soon as the artillery is compelled to cease fire because of the proximity of the assault troops.

Men armed with wire cutters cut a lane through the wire obstacle, hidden by the smoke screen. As an alternative measure, men with tube charges go forward and push their charges under the wire. These tube charges, which are similar to Bangalore torpedoes, contain 18 to 20 pounds of explosive. When the charges are in place, a designated engineer calls out "Ready for ignition!" (*Fertig zum zünden*), whereupon the commander of the obstacle-blasting party replies "All together, ignite!" (*All zusammen zünden*). The engineer then ignites the fuze and calls out "Burning!" (*Brennt*) to warn personnel nearby to get under cover. The explosion of the tube charge opens a lane in the wire. The engineers nearest the lane then shout "Gap here!" (*Hier Gasse*).

Besides blinding the defenders of a pillbox by means of smoke, the Germans also fire antitank guns directly at the embrasures of the pillbox.

(It seems highly probable that the shouting drill has been developed to enable the engineers to keep in touch with each other when visibility is poor or zero, and because of the difficulty of commanding the whole operation from a central command post.)

The flame-throwing detachment, having advanced with the remainder of the assault party from crater to crater, now moves through the gap in the wire and attempts to reach a point 5 or 6 yards from the pillbox.

Now that artillery fire has lifted from the area around the pillbox, the task of keeping the defenders' heads down is taken over by covering machine guns. The flame-thrower operators direct jets of flame at the various embrasures in the pillbox, in accordance with orders given before the operation began. The blinding effect of the jets enables the men with the pole charges to advance. When the flame-throwing detachment is about to run out of fuel, a designated engineer shouts "Last jet!" (*Letzter Strahl*). Each man who is carrying a pole charge advances to an embrasure and detonates his charge inside it. Prisoners state that these charges are effective even against closed embrasures.

If the pillbox continues to hold out, either of two alternatives is possible: (1) The engineers may throw smoke candles into the pillbox to drive out the occupants.

(2) The engineers may blow in the roof, using a charge weighing about 110 pounds. This charge, which may be carried in two pieces, is fitted with handles for easy transport. It is circular, and has a concave undersurface and convex upper surface. It is said to be about 10 inches thick in the center, but thinner toward the edge. Since the charge is constructed on

the hollow-charge principle, it can penetrate normal concrete or armor. It is detonated by a friction igniter.

As soon as an important pillbox has been taken, a swastika flag is draped over it as warning to friendly dive bombers. A pillbox in a fortress, for example, is considered especially "important."

3. ATTACKS ON TRENCHES

German engineers who have taken part in exercises involving attacks on trenches state that they have used ordinary assault methods, preceded by a liberal use of hand grenades.

For this purpose, certain men are trained as short-distance throwers (*Nahwerfer*) or as long-distance throwers (*Weitwerfer*). The flame-throwing detachments move directly behind the hand-grenade throwers, and the whole party is covered by machine-gun fire from the flanks.

[NOTE.—The Germans, having devised these tactics, are thoroughly familiar with the methods of defense against them, one of the most important of which is the use of pressure and trip antipersonnel devices in the vicinity of the dead angles of bunkers. Extremely meticulous intelligence is an essential for this type of assault.]

Explosives, Mines, and Demolitions

Advanced use of demolitions was one of the more specialist roles of the U.S. Army engineers. Using a variety of demolition charge types and weights, plus task-specific tools such as bangalore torpedoes and cratering charges, the engineers' explosive expertise would be put to use both in combat and in civil works contexts. In offensive action, combat engineer troops would use explosives mainly in the assault of fixed enemy positions, particularly protected strongpoints. A common tactic, for example, would be for a mixed engineer/infantry unit to close on an enemy bunker then post demolition charges through observation or firing slits, either throwing them in or pushing them through at a safe distance on the end of long poles. Alternatively, the engineers might utilize heavy shaped charges to cut directly through reinforced structures. Demolitions could also be used against armored vehicles, particularly to blow off tracks, although antitank mines were superior devices in this regard. Other common demolition duties of engineers included blowing up bridges and significant enemy structures, destroying obstacles preventing the movement of friendly armor or vehicles (a role most famously performed by U.S. Army engineers on the landing beaches of Normandy in June 1944), laying booby traps for personnel or armor, or disabling enemy equipment (e.g. blowing artillery gun barrels).

But explosives had far wider engineering applications than just combat. They might be used as a rapid cutting method for felling trees for timber. A considered placement of explosive packs and detcord could create drainage sumps and channels in milliseconds. Explosive charges could also assist in digging field fortifications in frozen or hard ground, or in clearing terrain features to improve fields of fire.

The "Explosives and Demolitions" chapter of FM 21-105, Engineer Soldier's Handbook, *provides a useful overview of the types, tactics, and applications of demolitions in the hands of a well-trained engineer. Underwriting almost every paragraph is a steely requirement that the engineers follow strict safety protocols when handling explosives, even those with wider safety parameters. At the same time, the job of using demolitions required a nuanced understanding of the relationship between type, weight, and configuration of explosives, the material effect, and the tactical purpose. The manual issues an unflinching directive in this regard: "You must not fail."*

From FM 21–105, *Basic Field Manual: Engineer Soldier's Handbook* (1943)

CHAPTER 6
EXPLOSIVES AND DEMOLITIONS

■ 45. The Job of Destruction.—*a.* As an engineer soldier, one of your most important jobs is the handling of explosives and demolition tools. It takes training to become an expert demolition man. There is a great deal to learn. In this chapter you will find enough fundamentals to give you a good start. With these essentials and with experience you can gradually become an expert.

b. You must learn this job thoroughly. It is a great responsibility. When you are given the job of blowing up a bridge, a road, or a building, that bridge, or road, or building *must be destroyed at the specified time.* There can be no mistakes. Demolitions are usually ordered at critical times; and the failure of a single demolition may cost the lives of hundreds of men. You must not fail.

■ 46. Equipment.—Demolition sets are issued to all general engineer units and many special engineer units. Each set includes a supply of explosives and the necessary tools and equipment for preparing, priming, and firing demolition charges. Earth–drilling tools, wood augers, and rock drills required for placing charges are available in pioneer and carpenter sets.

■ 47. Explosives.—*a. TNT.*—(1) TNT (trinitrotoluene) is the standard explosive for Army use. It is issued in ½-pound blocks encased in a cardboard container closed at both ends with lacquered tin. One end of each block has a cylindrical hole, approximately ⁵⁄₁₆ inch in diameter and 2⅛ inches long, for receiving the cap. TNT is one of the safest explosives to handle, if you know how to use it. It is insensitive to shock and will not detonate even under strong pressure or severe blows. It requires the special issue cap or detonating cord to set it off. In small quantities it can be burned without danger of detonation, but in large quantities the heat generated will raise the temperature to the detonating point.

(2) TNT will not dissolve in water and hence is suitable for underwater demolition work.

b. Nitrostarch.—Nitrostarch is issued in ½-pound, cardboard-covered blocks of the same size and shape as TNT, and in 1-pound paper-wrapped packages. Each of the 1-pound packages is made up of four ¼-pound packages, which, in turn, are made up of three ½-pound blocks. Each of these blocks has a cap hole extending all the way through it. Nitrostarch is similar in many respects to TNT.

c. Dynamite.—Dynamite is issued in approximately ½-pound sticks, approximately 1¼ inches in diameter and 8 inches in length. Fifty percent straight dynamite is equal in strength (pound for pound) to TNT. It is much more sensitive than TNT and may be detonated by a blow with a metal instrument, or by flying sparks struck from metal striking metal. When frozen it is especially dangerous and must be handled with extreme care.

d. Ammonium nitrate cratering explosive.—Ammonium nitrate cratering explosive is issued in 40-pound charges, each packed in a cylindrical container of tin or other moisture-proof material of equal strength. The metal container is about 8¼ inches in diameter and 17 inches in height; another type of container, made of waterproofed cardboard, is 7 inches in diameter and about 21 inches high. Two tubes are secured to the outside wall of each container, one for receiving the detonating cord, and the other the special cap. If exposed to air, ammonium nitrate explosive absorbs moisture rapidly; consequently, it must never be removed from the container. It is used principally in making crater obstacles for tanks and other motorized vehicles.

■ 48. Bangalore Torpedo.—*a.* The bangalore torpedo is a metal tube or pipe filled with explosives. Its primary uses are to cut gaps in barbed wire obstacles and to cause detonation of mines. The standard bangalore torpedo, about 2 inches in diameter, is issued in 5-foot watertight sections already filled with explosives. Sleeves are provided for connecting sections to extend torpedoes to any desired length. By fastening the rounded nose on the forward end, you can push the torpedo through a band of barbed wire without getting it caught on the wires.

b. To explode the torpedo, an electric or nonelectric cap, or primacord, is inserted in the cap well in the trailing end of the torpedo. When several sections are joined to form a long torpedo, it is necessary to place a cap only in the last section, since detonation of one section will cause the whole torpedo to explode. If standard-type torpedoes are not available, you can make bangalore torpedoes by filling a pipe (for example, a 2-inch water pipe or an old drain pipe) with explosives; the ends are closed with wooden plugs, and one end is primed by making a hole through one of the plugs; a primer made with TNT block and primacord is placed inside the torpedo and the primacord end is drawn through the hole in the wooden block.

c. Remember that each 5-foot section of the bangalore torpedo is loaded with about 10 pounds of high explosive, and the same precautions in handling and firing must be taken as when other military high explosives are used.

■ 49. Firing Materials.—*a. Caps.*—Caps are placed in charges to set them off. Standard commercial caps will not detonate TNT or ammonium nitrate cratering charge; therefore the army has adopted a special cap. Caps are classified as electric or nonelectric, depending on whether they are set off by electricity or fuze. Both types must be handled with great care, because they may be set off by dropping or hitting them, or exposing them to excessive heat.

b. Exploders.—Exploders are used to supply electric current to set off electric caps. The 10-cap exploder is operated by a quick twist of the handle. The 30-cap exploder is operated by slowly pulling the handle all the way up and then pushing it all the way down as fast and as hard as possible.

c. Firing wire.—Firing wire, carried on a metal reel, is used to connect the exploder to wires of electric caps placed in charges. It is issued in 500-foot lengths so that a man may fire the charge from a safe distance. Cap wires are connected to the free end of the firing wire, and the exploder is connected to the end which is fixed to the metal reel. When extremely large charges or steel-cutting charges are being fired, two or more reels of wire may be connected so as to enable the firer to fire the charge from a distance of 1,000 feet or more.

d. Time fuze.—A time fuze is used to set off nonelectric caps. It consists of a train of black powder contained in a waterproofed textile covering which may be either white or orange. When using a time fuze, cut it to the desired length and crimp one end in the nonelectric cap. Light the other end with a match or fuze lighter after the explosive charge has been prepared. Always be sure to use a fuze long enough to enable you to reach a place of safety before the charge explodes. A time fuze burns at the rate of about 2 feet per minute.

e. Fuze lighter.—The fuze lighter is used to light a time fuze. The open end is placed over the end of the time fuze where it is held in place by means of teeth inside the fuze lighter. These teeth permit the fuze to enter, but are inclined so as to bind the fuze and prevent its removal. It is unnecessary to crimp the lighter. Pulling the handle causes a flame inside the lighter which lights the fuze even in wet or windy weather, if the lighter and the powder train in the fuze have been kept dry. The fuze lighter should be set off by means of a steady pull (not a jerk).

f. Detonating cord.—A detonating cord consists of a train of high explosive contained in a waterproofed textile covering. It is set off by a cap taped or tied to it. Instead of burning like a time fuze, it explodes like other high explosives and will set off other explosives properly connected to it. It is used mainly to set off a number of charges at one time or to fire a charge in a deep hole. Its action is instantaneous; therefore, whether the detonating cord is fired by an electric or nonelectric cap, the firer should take the same precautions as if the cap were placed directly in the charge.

g. *Crimper.*—The crimper is used to crimp the open end of the nonelectric cap around the time fuze and to cut time fuze. One leg of the handle is pointed for use in making holes for caps in dynamite, and the other leg has a screw-driver end.

h. *Tools for boring holes.*—Holes in earth, concrete, rock, or other material in which explosive charges may be placed are made with many kinds of hand and power tools. [. . .] Tools [. . .] include the air compressor, with its rock and pavement-breaking attachments, and two types of earth auger, one of which drills holes 6 inches in diameter and the other holes 10 inches in diameter. These augers are useful in drilling holes to place charges for road craters.

■ 50. PRIMERS.—A primer is a block or package of high explosive with a cap placed in it, or detonating cord tied through or around it. A charge is primed by placing in it a primer prepared as above. The purpose of the primer is to insure detonation of charge. The cap or detonating cord sets off the block or stick of explosive in which it is placed, which in turn sets off the remainder of the charge.

a. *Primer of TNT or dynamite, with nonelectric cap and fuze.*—Cut fuze off *square* with fuze cutter. See that no dirt is in the open end of the cap, then slip it over the fuze as far down as it will go. Hold the fuze in one hand and crimp cap to fuze with cap crimper, making crimp near open end of cap. Next, put cap with fuze attached into the hole in TNT block or dynamite. The twine holding the cap in place must be tied so that when the fuze is pulled, the twine takes the pull and the cap is not moved in the hole.

b. *Primer of TNT block or dynamite, with electric cap.*—Place cap in hole in TNT block, making a clove hitch in the cap end of the lead wires. Pass this loop around TNT block and pull tight, leaving the wire slack from the loop to the cap. The same procedure is used with a dynamite stick; another method is to tie a preliminary half hitch near the bottom of the stick before tying the clove hitch near the top.

c. Primer with detonating cord and TNT block.—Tie cord directly to the TNT block. To do this tie one-half of a clove hitch, take one full turn around the block, then tie the second half of the clove hitch. Pull all three turns tight so they will fit snugly against each other and against the block.

d. Primer with ammonium nitrate cratering charge.—For electrical firing place an electric cap in the cap well provided on the side of the container. Make several turns with the lead wires around the knob above the cap well and pull tight, so that a pull on the lead wires will not dislodge the cap. To make a primer using detonating cord, pass the detonating cord all the way through the tunnel (from top to bottom) on the side of the container and tie an overhand knot in the lower end to prevent the cord from pulling out of the tunnel. Both means of detonating should be used simultaneously. Do not use detonating cord or electric-cap leads to lower the ammonium nitrate cratering charges into holes; use a cord attached to the ring in the top of the charge.

[. . .]

■ 54. PLACING CHARGES.—The officer, or noncommissioned officer, in charge of each demolition project gives definite instructions as to the sizes of charges to be used and where and how they are to be placed. Failure to use the proper amount of explosive results in failure of the demolition project, and placing a charge incorrectly may be just as disastrous. Don't try to save yourself work by using a smaller charge or by placing the charge in a location that is easier to get to than the location directed. If, for any reason, it is impossible to place the charge in the location or manner directed, report this fact to the officer or noncommissioned officer in charge. The following points will be helpful in enabling you to place charges properly to attack concrete, steel, and timber.

a. Crater.—Figure 103 illustrates how to prepare a borehole to blow a crater with TNT. The depth of the hole is determined by the officer in charge. In place of the TNT blocks, one or more ammonium nitrate cratering charges may be used.

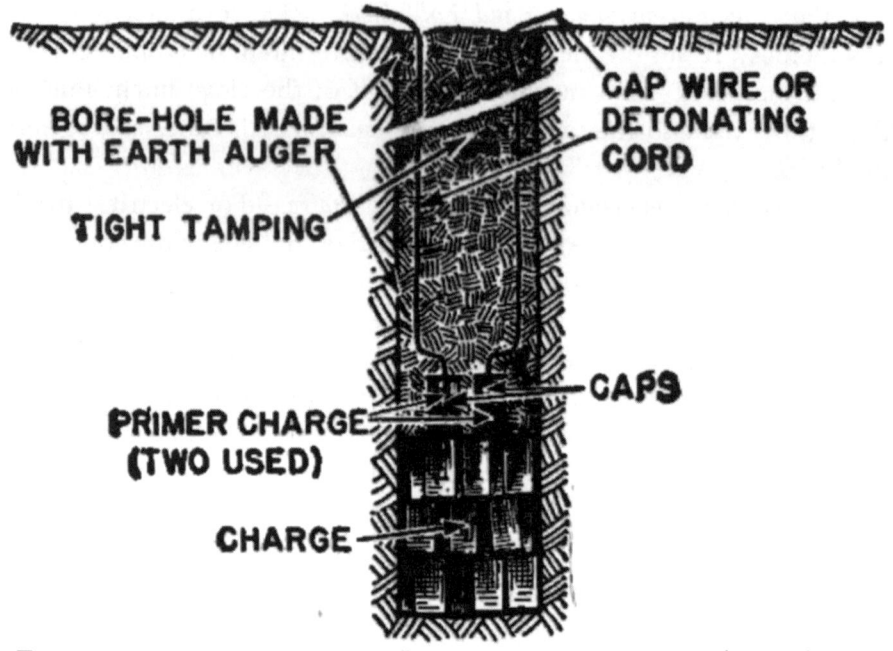

BORE-HOLE MADE WITH EARTH AUGER

TIGHT TAMPING

CAP WIRE OR DETONATING CORD

CAPS

PRIMER CHARGE (TWO USED)

CHARGE

FIGURE 103.—Bore hole loaded to blow crater.

b. Concrete.—Because of the difficulty of placing internal charges, concrete is normally attacked by external charges calculated to blast through its entire thickness. In attacking a concrete wall, or a bridge abutment or pier, the entire amount of explosive is normally concentrated at the midpoint of the structure, if the width is not more than twice the thickness. Where the width is more than twice the thickness, two or more charges are used, each charge being large enough to shatter the thickness of the wall, and the distance between successive charges being not greater than twice the thickness of the structure.

c. Steel.—The effects of an explosion are very localized in steel. Only that portion of steel which is in close contact with the explosive charge is cut. A concentrated charge placed on a steel plate will simply blow a hole in the plate or dent it directly under the charge. Therefore, if it is desired to cut a steel plate, the explosive charge must be distributed over the entire line along which the cut is desired. Likewise in cutting steel I-beams, built-up girders, columns, etc., the charge must be distributed so that the entire cross section of the member will be cut. Charges must never be placed directly opposite one another on opposite sides of a steel plate or members. When charges must be placed on opposite sides they should be 3 inches apart along the member.

d. Timber.—External charges for cutting trees, round timbers, etc., are placed around one side of the object. The charge should not extend more than halfway around the object being cut. External charges for cutting square timber members are placed on one face along the line of the desired cut. In the case of rectangular timbers not having a square cross section, the charge is placed on one of the faces having the longer dimension. For example, to cut 10- by 12-inch timber, the charge is placed on one of the 12-inch faces. Internal charges require only about one-sixth as much explosive to cut timber as external charges. Hence, when time permits, an internal charge is used. Such a charge is placed in a bore hole in the timber, and well tamped with moist clay or mud. An underwater charge for cutting a timber pile is fastened to a board, then shoved beneath the water surface in close contact with the pile, where it is lashed or nailed in place.

■ 55. SAFETY PRECAUTIONS.—*a.* Don't forget that explosives are always dangerous.

b. Don't smoke while handling explosives, nor handle explosives near open lights, fires, or stoves.

c. Don't handle or keep explosives in or near places where there are large groups of people.

d. Don't open cases of explosives near caps or other explosives.

e. Don't leave explosives in the open where they may be stolen, tripped over, or where animals can get at them.

f. Don't leave explosives in wet or damp places; keep them locked up in a dry place.

g. Don't use frozen, chilled, or bleeding (leaking) dynamite.

h. Don't try to thaw dynamite; have an expert do it with the proper equipment or get some other explosive.

i. Don't put dynamite near steam or hot water pipes or stoves.

j. Don't keep or transport caps anywhere near explosives.

k. Don't drop or tap caps or carry them in your pocket.

l. Don't take caps from box with wires or nails; use fingers only.

m. Don't leave caps out in the sunlight, or where they may be stepped on or run over. Keep them in their box until time to use them.

n. Don't pull on wires of an electric cap.

o. Don't hold caps in hand while crimping; place cap on one end of fuze and hold fuze end.

p. Don't crimp cap with anything except issue cap-crimper.

q. Don't tamp with iron or steel bars or tools. Use only blunt wooden tamping stick and tamp lightly at first, then harder.

r. Don't force primer into a drill hole; make hole big enough.

s. Don't cut fuze too short; explosion may occur before safe distance can be gained.

t. Don't risk misfire by using too weak cap; employ proper cap for explosive used. You must use special issue cap to detonate TNT.

u. Don't explode charge until everyone is safely under cover or out of danger.

v. Don't connect firing wires to exploder until ready to fire charge.

w. Don't spring (enlarge) bore hole and then immediately reload; the bore hole will still be hot and may explode charge.

x. If charge fails to explode, wait at least 30 minutes before investigating it, unless an officer or experienced demolitions man directs otherwise. Explode misfired charge by means of another charge placed as close as possible to misfire. The old charge should not be disturbed.

Engineer assault on D-Day

The Allied landings on June 6, 1944, in Normandy, France presented a formidable challenge to U.S. engineers. In their designated beach sectors—Omaha and Utah—the landing forces faced heavily defended elements of the German Atlantikwall, substantial ferro-concrete fortifications overlooking the beaches and bristling with firepower. But before they could tackle those, the immediate priority was to clear various emplaced obstacles on the shoreline, specifically those designed to prevent troops successful reaching and consolidating the beachhead. The obstacles included wooden posts with mines affixed to the top, 'hedgehog' frames of steel beams to trap vehicles and landing craft, and 'Element C' wooden and metal boat barriers.

The following text comes from a post-D-Day edition of Combat Lessons. *Drawing on reports and first-hand testimonies, it collects some of the engineering lessons from that fateful day:*

Some Experiences in Gapping Beach Obstacles

Officers of a *Provisional Engineer Group* which assisted in the D-day landing in NORMANDY give this account of their experiences and draw several conclusions that should be considered for similar missions in the future:

—**Mission** "The mission of the group was to blow sixteen 50-yard gaps through all obstacles within the tidal range of the selected objectives, and later to widen and extend these and clear the entire beach area of obstacles."

—**Organization and Plan** "The force included two Engineer Combat Battalions, 10 tank dozers, and 21 naval combat-demolition teams. Gap-assault teams were organized, each composed of one navy combat-demolition team subdivided into two mine crews and two demolition crews. The plan was for these engineer units to land just after the infantry; the infantry was to work its way through the obstacles, leaving the task of gapping to the engineers behind them.

"Each gap-assault team was to land at a designated point, prepare a 50-yard gap, and mark it from low- to high-water lines. Support and command teams were to land 5 minutes later to assist the gap assault teams. Their equipment, all hand-carried, was especially prepared and waterproofed in advance."

—**Obstacles Encountered** "Boat teams 7 and 8 encountered the following types of obstacles in the order listed: a line of posts interspersed with log ramps; consecutive bands of log ramps; another line of posts and ramps; and a line of hedgehogs.

"Boat team 6 encountered first a great deal of 'Element C' and then staggered rows of hedgehogs. The hedgehogs differed from the type expected. They were of a lighter material, but bolted together and reinforced in the middle; three sticks of dynamite were not enough to reduce them. Other teams found different patterns of obstacles."

—**Divergences from the Plan** "Some of the teams landed simultaneously with or ahead of the infantry, the others very shortly after the infantry. Gapping operations of these crews were seriously impeded by infantrymen who took cover behind or near the obstacles that were to be blown.

—**Some Teams' Experiences** "Boat team 8 landed on time but at a point to the left of its designated landing. The infantry had not preceded them, so these engineers were the first on the beach. Hostile fire was light until the men were actually on the beach; at that time the enemy opened up and the engineers dropped to the sand and dispersed. However, each man tied a charge to the obstacle near which he had taken cover, using his judgment as to the amount of explosive to use. (In most cases, one charge was used on the straight posts, three on the log ramps.) These charges were blown before the infantry landed. On the second tide, the infantry had moved ahead and the hedgehogs were blown; these required one to three charges each.

"Boat team 6 touched down 50 to 100 yards to the right of its objective a few minutes after H-hour. There were no infantry on the beach; enemy machine- gun fire was heavy. However, the charges were ready to explode 10 to 15 minutes after the work started. The first infantry landed immediately after the gap was blown and swarmed through. As soon as they were through, a second series of charges was set off. A complete gap was blown through in about 30 minutes.

"On the second tide, this team went out to widen the gap. The infantry were piling up and milling around; it was necessary to resort to smaller and fewer charges to avoid injuring them."

Lessons Learned in Normandy

"1. When Engineer teams are landed close behind the infantry they must expect interference with their work until the infantry moves inland.

"2. It developed in this case that if the mines had been removed, the initial landings would have been just as successful without any demolition work. Mines did most of the damage that was done. The LCTs smashed right through the other obstacles.

"3. Demolition squads which are landed well to seaward of the outermost obstacles are able to accomplish more before the tide interferes with the work.

"4. In landing operations, demolition teams must expect and allow for many unforeseen difficulties which will delay their work."

★★★

Don't get killed by Mines and Booby Traps *was a U.S. War Department publication issued throughout the U.S. armed forces from November 1944. Its requirement was pressing. As U.S. units pushed Axis forces back towards their homelands, the retreating and bitter enemy became ever more inventive in leaving behind booby traps and mines for the unwary, both to inflict casualties and to slow*

On a training exercise in England in April 1943, men of the 112th Engineer Regiment practice a flamethrower assault on a pillbox. (Signal Corps Archive)

the advance. The devices reflected the remorseless ingenuity of the enemy while also taking advantage of the carelessness of tired and curious soldiers operating in places where they thought they were safe. German booby traps, for example, would be hidden in or under almost any space or object large enough to hold an explosive device—a tin can, the pipe of a water pump, a cooking pot with the lid on, a warped floorboard, above a door, under the body of a dead soldier. Some devices, when detonated, would only kill or wound the soldier in the immediate vicinity. Others would trigger huge webs of interconnected charges, bringing down entire housing blocks. Mines and booby traps account for an estimated 3 percent of U.S. deaths and 4 percent of U.S. injuries during World War II.

Given their expertise in ordnance, the U.S. Army engineers were used intensively for mine and booby trap clearance during the conflict. "Call the engineers" became a common refrain when U.S. infantry encountered either a minefield or suspect objects. Don't get killed by Mines and Booby Traps *is a useful document from an engineer's point of view because it was taken from the engineer manual FM 5-31,* Land Mines and Booby Traps, *and thus provides direct insight into engineering techniques of detecting and defusing both Japanese and German devices. We should remember that the engineers were also responsible for laying U.S. minefields and setting their own booby traps—thus they brought to the task their own understanding of how to employ explosive devices most effectively.*

★★★

From *Don't get killed by Mines and Booby Traps* (1944)

CHAPTER ONE
WHY LEARN ABOUT MINES AND BOOBY TRAPS?
—because they <u>KILL!</u>

Mines and booby traps are not placed by magic; they are placed by the enemy or our own troops. They were once safe to handle, and they are always made unsafe by somebody doing something to them—removing the safety pin or compressing and latching a spring. A soldier who has had a little experience with mines can always find a way to return them to their original, safe condition.

Veterans returning from overseas say that all soldiers (yes, even WACs) should be taught how mines work, how to identify them, and what measures to take against them.

This pamphlet is to give you something to read and study before going into territory previously occupied by the enemy. It will acquaint you with various types of mines used, how they are used, where they are used, and what to do about them.

[. . .]

CHAPTER TWO
WHAT ARE MINES?
They are hidden __DANGER!__

ANTITANK MINES are explosives that STOP VEHICLES

ANTIPERSONNEL MINES and BOOBY TRAPS STOP PEOPLE

A **booby trap** is an explosive charge arranged so any disturbance of a seemingly harmless object sets it off. Booby traps may be prepared charges or antipersonnel mines and are used to delay, demoralize, and produce casualties.

The booby trap differs from the antipersonnel mine only in the employment by the enemy. Antipersonnel mines serve a tactical use while booby traps are used principally to scare, harass, and demoralize all our troops in captured territory. The booby trap can be quickly constructed and set up in any number of ways limited only by the ingenuity of the person setting the trap. The enemy has booby trapped practically everything including their own dead and even tombstones on our dead. The enemy has used almost every known type of ordnance equipment for booby trapping including land mines, grenades, aerial bombs, artillery shells, and weapons. The enemy preys especially on the souvenir hunter. Some ingenious booby traps include double bottom trunk, tobacco tins, parasols, ping pong balls, pistol disguised as a cane, pistol disguised as a fountain pen, devices using flashlights, devices using a pipe and devices using matchboxes. All enemy ordnance should be left entirely alone, except for marking its location and reporting it to your commander.

CHAPTER THREE
WHAT SETS THEM OFF?
__YOU__ DO!

HERE'S *HOW*. . .

A mine or booby trap is set off by a fuze. When an outside force acts on the fuze, it fires the explosive in the mine. YOU apply this force in the following ways:

you step on 'em
you drive over 'em

. . . and set off a pressure fuze. The pressure causes a striker pin to hit a percussion cap; this causes the cap to go off, exploding the mine.

YOU *PULL* THINGS
. . . and set off a pull fuse. There are two common types of pull fuses.

It may be a **PERCUSSION** fuze. The pull on the wire releases a spring-driven striker pin which hits and fires a percussion cap.

It may be a **FRICTION** fuze. A pull creates friction (like striking a match), causing a flash which fires a cap.

YOU *LIFT* THINGS
. . . and set off a **PRESSURE-RELEASE** fuze. Taking the weight off a release plate causes a spring-driven striker to hit and fire a percussion cap.

YOU *CUT* THINGS
. . . and set off a PULL or **TENSION-RELEASE** fuze. A striker pin held back by a taut wire is released when the wire is cut or pulled, setting off a percussion cap.

YOU *MOVE* THINGS
. . . and complete **ELECTRIC CIRCUIT** to fire an electric cap, setting off main charge.

CHAPTER FOUR
WHAT DO THEY LOOK LIKE?
here are a few—

U. S. ANTITANK MINE M1A1

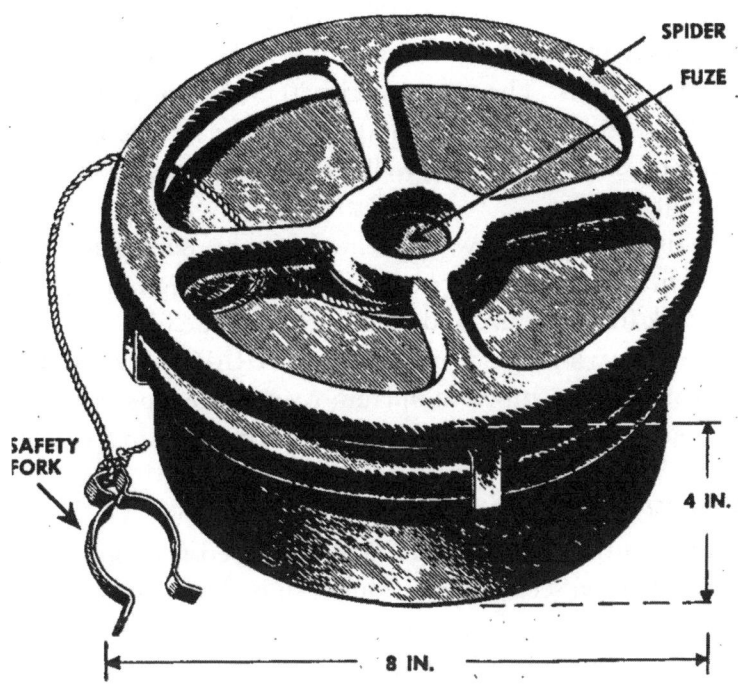

This mine is the **standard US antitank** mine. It weighs about 11 pounds, of which 6 pounds are TNT. A pressure of 500 pounds on the fuze or 250 pounds on the edge of the spider fires the mine.

To assemble the mine, remove spider and place fuze in cavity of mine body. Hook the two legs of spider under rim, pass other two legs through notches, and turn spider one-eighth of a turn.

To lay and bury the mine, first remove safety fork and place mine in a cover. Then place mine in hole and fill in so top of spider is at least one-quarter inch above original ground level. Replace sod and complete camouflage. If mine is buried with spider down, upper surface should be not more than 1 inch below ground surface.

To disarm the mine, cut away cover to reveal fuze; then, if fuze is not damaged, replace safety fork, check for booby traps and lift mine. If safety fork does not go on easily, do not force it or remove mine by hand. Instead, attach a 50-yard length of rope or wire, drag mine to safe place, and destroy with explosive.

U. S. HEAVY ANTITANK MINE M6

The **American heavy antitank mine** is the answer to the enemy's use of the heavy tank. The mine weighs 20 pounds of which 12 pounds is explosive. A weight of 300 to 400 pounds on the pressure plate fires the mine. There is a booby-trap well on the side and one on the bottom for anti-lifting devices.

To arm the mine, unscrew and remove the pressure plug on top and inspect the fuze well to make certain it is free of foreign matter. Remove the safety fork from the fuze and then insert it in the fuze cavity. Replace the pressure plug with the side up that reads, **ARMED, THIS SIDE UP**.

To disarm the mine, unscrew and remove pressure plug, then withdraw fuze and replace safety fork on fuze. Carry mine and fuzes separately.

U. S. LIGHT ANTITANK MINE M7

The **light antitank mine M7** was developed for hasty mine fields laid to provide local security. The mine can be lifted and relaid as often as necessary. The mine is rectangular in shape, weighs 4½ pounds, and contains about 3 pounds of explosive. The fuze is the same as for the M6 mine, and a pressure of 150 to 250 pounds will set the mine off.

The M7 is laid with its long side across the expected direction of attack. To be effective against heavy tanks, the mines should be laid double, one on top of the other. To arm the mine, lift pressure plate and insert fuze, first making sure well is clear and free of foreign matter; then remove safety fork. Avoiding downward pressure, slide pressure plate into position. Center it over fuze with rivets on either side of mine in vertical slots of pressure plate.

When burying the mine, place it in a cloth cover and bury so surface of pressure plate is not more than 1 inch below ground.

The mine has a booby-trap well on one end.

To disarm the mine, carefully search for booby traps, lift mine, and replace safety fork.

GERMAN TELLERMINE 35

The Germans have developed mine warfare to the greatest extent of any nation. The most common of their antitank mines is the **Tellermine** named after the German word **plate**. There are four types of Tellermines, each containing about 12 pounds of explosive and each weighing about 20 pounds.

All Tellermines require about 250 to 400 pounds pressure to set them off.

Also each mine has a booby-trap well on the side and bottom.

The original Tellermine, known as **TMi 35**, was designed, as the number implies, in 1935. It was used extensively in Europe during the 1939–1940 campaign. It has been used since then, but not as frequently as the later models.

The fuze is the brass nob on the top. It has two safeties: one that requires a coin or screw driver to turn a disk on the knob to "Scharf" (armed) or "Sicher" (safe); the other a bolt projecting from one side of the fuze.

To arm the mine, the disk on top of the fuze is turned to "Scharf" and the safety bolt is pulled out to the side.

To disarm the mine, push the projecting safety bolt in gently. Do not force. This makes the mine safe. It is not necessary to turn the disk on top the fuze to "Sicher," as this is only secondary safety used when transporting the mine.

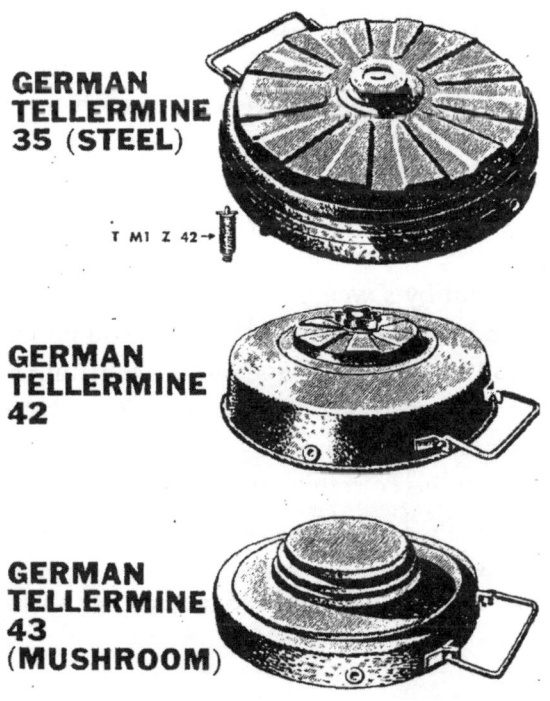

GERMAN
TELLERMINE
35 (STEEL)

T MI Z 42→

GERMAN
TELLERMINE
42

GERMAN
TELLERMINE
43
(MUSHROOM)

The three later models of the Tellermine, known as the TMi 35 (Steel), TMi 42, and TMi 43 (Mushroom), have been most frequently used by the Germans.

These mines are about the same size and weight as the original model, and also have a booby-trap well on the side and on the bottom.

All three mines use the TMiZ 42 or TMiZ 43 fuze. Only the TMi 35 (Steel) can use the same fuze as the original Tellermine with a minor modification.

With the development of the TMiZ 43 fuze, it is no longer possible to disarm these mines by removing the fuze. The TMiZ 43 fuze is similar to TMiZ 42 except that when it is placed in the mine and the pressure plug is screwed on, a secondary shear pin is broken so that upon removal of the pressure plug the mine explodes.

These mines can be safely destroyed in place or pulled out to a convenient place with a 50-yard length of cable or rope and then destroyed.

WOODEN BOX MINE 42 (HOLZMINE)

In Sicily and in Europe, the **Holzmine 42** is being used. It is commonly called the German wooden box mine. We can expect to find it in large quantities as it contains no critical materials and is easy to construct. The mine weighs 18 pounds, of which 11½ pounds are explosives. There are enough nails and wire hooks in the mine so the mine detector can locate it. The fuze in this mine is the ZZ42 or bakelite fuze, which is a common fuze used with antipersonnel mines and booby traps.

The mine is set off by a weight of 200 pounds or more moving over the pressure block. This causes the block to move downward, breaking the retaining wooden dowels, pushing out the actuating pin of the ZZ42 fuze, and setting the mine off. Booby-trap wells can be easily placed on the side and bottom of the mine.

To disarm this mine, carefully unfasten and remove the lid. Lift pressure block clear of shear flange, rotate 180°, place it so it bears on the supporting block.

This mine is tricky; only trained personnel should attempt to defuze it.

GERMAN 'RIEGEL' MINE 43

The newest German antitank mine is the RMi43, commonly called **Riegel mine 43** or Sprengriegel 43 (Spr R43). The mine has three main parts: (1) an encased charge of TNT contained in (2) a sheet-steel tray, and (3) a lid which fits over the tray and acts as a pressure plate on the charge. The total weight of the mine is about 20 pounds, of which 9 pounds is TNT. The mine is light khaki in color. Two fuzes ZZ42 are used, one at either end of the mine.

The mine is fired either by (1) enough pressure on the lid to shear one or both of the shear wires, (2) by the tilt fuze 43 (KiZ 43) or the functioning of anti-lifting or trip-wire fuzes fitted in the five sockets provided, or by (3) remote electric control.

THIS IS AT ALL TIMES A DANGEROUS MINE TO DISARM— IT MAY BE DANGEROUS TO HANDLE IN ANY WAY—IT SHOULD ALWAYS BE DESTROYED IN PLACE.

Note: Tilt fuze 43 (KiZ 43) is like a toggle switch. The fuze is inserted in the well on top of the Riegel mine and has an antenna 2 feet long sticking up in the air. A pressure of 1½ pounds on the antenna in any direction will set the mine off. The fuze can be used with other types of mines by burying the mine upside down and placing the antenna in the booby-trap well.

JAPANESE ANTITANK MINE TYPE 93

The **Japanese antitank mine 93** is a small mine weighing only 3 pounds, of which 2 pounds is explosive. It has a tin shell and is painted olive drab. To be effective against tanks, the mines must be used in groups of three and four. The mine has no booby-trap wells. The mine can be used with either of two fuzes. One fuze will set off the mine with 70 pounds pressure, the other with 250 pounds pressure.

To disarm the mine, unscrew the pressure plug and carefully unscrew the whole fuze and lift it out. If the brass safety cap is available, screw it firmly into the top of the fuze before removing the fuze.

These mines have been found buried upside down with additional explosives placed beneath them to increase their effect.

We have not met the Japanese on any terrain suitable for large-scale tank warfare. Thus they have not employed as many antitank mines as have the Germans. But as we advance nearer the Jap homeland, we can expect them to use mines more and more.

JAPANESE BEACH MINES

JAPANESE BEACH MINES

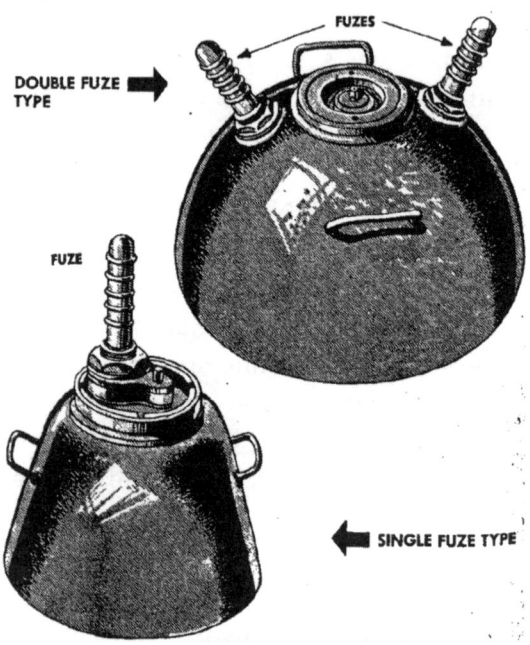

During our island hopping in the Pacific, our Army and Marine Corps have run into the Japanese antiboat mine. These mines have been found between the fringing reefs and the high-water marks on the beaches and are designed to destroy landing craft. They have also been used in conjunction with underwater obstacles, steel wires being fastened between the horns and obstacles to act like trip wires.

The two types of beach mines are known as the single-horn beach mine and the double-horn beach mine. They both use inter-changeable horns as fuzes, which when bent or broken set off the mines. A push or pull of about 200 pounds on the horn is necessary to break the glass vials in the horns. The double-horn mine weighs 106 pounds of which 46 pounds is explosive; the single-horn mine weighs 66 pounds of which 22 pounds is explosives.

JAPANESE 'YARDSTICK' MINE

This mine is called the "Yardstick" mine because it is 36 inches long. It is primarily an "anti-vehicle" mine and contains four fuzes, or "pressure points" distributed along its length, hence covering more area than any earlier Japanese mine. The mine contains eight ¾-pound blocks of explosive. One end of each block is molded to fit a fuze. Two blocks placed with molded ends together completely enclose one fuze with exception of release plunger, which protrudes from upper surface. Four two-block units placed end-to-end fill the case. A common safety wire through one end of the case passes through all four fuzes. To arm the mine, this safety wire is pulled out from one end.

To defuze the mine, first examine for booby-trapping and then lift mine. Remove both end caps and, gently pushing on explosive block at one end, force charge and fuze through opposite end. Do not allow fuze to drop. Place a short piece of #16 wire or small nail through safety pin hole of each fuze. If mine case is deformed, detonate mine in place by explosive.

U. S. ANTIPERSONNEL MINE M2A3

The **American antipersonnel mine M2A3** is of the bounding type and when actuated by any of several methods projects a mortar-like shell about 6 feet into the air, where it explodes. It is more deadly than a 60-mm mortar because it explodes above ground, thereby producing more casualties in a larger area.

The mine has a tube containing the propelling charge and a fuzed shell, and a small pipe to which the primer and fuze assembly are screwed. It stands on a base plate to which the tube and pipe are welded. The fuze is the combination pull-and-pressure type, requiring a pull of 3 to 6 pounds on the pull ring or a pressure of about 20 pounds on the pressure cap to set it off.

To lay and arm the mine, screw the fuze onto the mine, making sure the safety screw and safety pin are in place; then place the mine in a hole on a firm foundation and fasten the trip wires. Remove locking screw and carefully remove safety cotter pin. If the safety pin does not come out easily, do not force it; it is likely that the striker is released, in which event removing the cotter pin will set the mine off prematurely.

To disarm the mine, insert safety pin in fuze and screw in locking screw. Disconnect trip wires, check for booby traps, and lift mine.

GERMAN "S" MINE

Everyone has heard of the **Bouncing Betty, Silent Soldier**, and **The Jumping Jack**, all nicknames for the **German "S" mine** or SMi35.

The **S mine** consists of two parts: (1) the 4-inch round outer case and (2) the "jumping" inner case, which when set off comes up out of the ground 3 to 6 feet and explodes, sending 350 ¾-inch-diameter steel balls flying in all directions. The mine can be used with pressure fuze, pull fuze, tension-release fuze, or with a combination of fuzes.

Here's how it works. When the fuze is set off in any of several ways, it sends a flash down the center tube, setting off a delay pellet. The propelling charge in the bottom throws the inner case upward about 3 to 6 feet, where it explodes and sends shrapnel flying in all directions.

To disarm the mine, carefully uncover it to identify the fuze or fuzes and insert safety pins in the safety holes of all fuzes. After checking both ends for additional fuzes, cut any trip wires.

GERMAN SCHÜ-MINE 42

The **Schü-mine** was originally designed to prevent detection by the mine detector and so had no metal in it. It depends on blast rather than shrapnel to produce casualties. It is laid where personnel will step on it, and the ½ pound block of explosive will injure the person stepping on it.

The hinged cover acts as a pressure plate and a downward pressure of from 6 to 11 pounds on the lid will cause the notched cover front to force out the actuating pin in the ZZ42 pull fuze and set off the mine. The fuze is the same as for the Holzmine and the Riegel mine 43, and is now being made of metal.

As stated previously, the fuze has no safety; therefore, great caution must be taken when disarming the mine. To disarm it, carefully lift the lid without exerting pressure and see whether the actuating pin of the fuze is still firmly in the striker. If not, destroy the mine in place with a small charge.

Before removing the mine, carefully check for any anti-lifting devices on or near the mine.

GERMAN BUTTERFLY BOMB

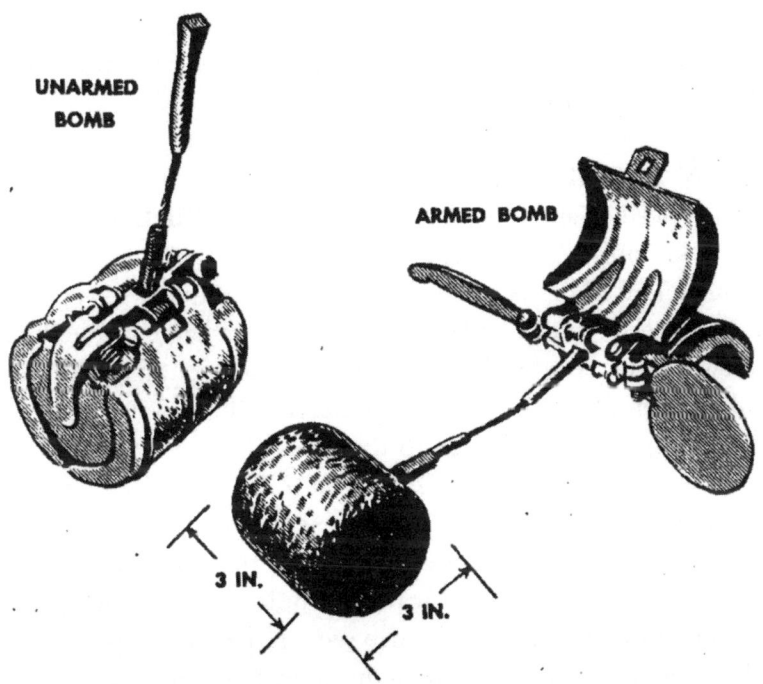

The German **Butterfly Bomb** is the favorite bomb used against personnel on beaches, in camps, on airfields, or wherever else troops are likely to assemble.

The bomb itself is 3 inches long and 3 inches in diameter, having sheet metal wings attached by a 5-inch length of wire. The bomb is red, yellow, or olive green, and the wings are yellow or olive green. These bombs can be fitted with one of four different types of fuze. Two of these are made so they will detonate the bomb either in mid-air or as it hits the ground, depending on their setting. The third is a delay-action fuze which will function at any time up to 30 minutes after falling. The fourth and last is an anti-handling fuze, which goes off instantly when someone bumps the mine or merely touches it. No bombs of this type should be approached for at least 30 minutes after they were dropped. Only in extreme urgency can this rule be broken. The best way to remove the menace is to place a small charge as close as possible and let the concussion set the bomb off. If near buildings, planes, or vehicles, carefully build sandbag walls around the bomb before exploding it.

One point to remember—place warning signs and call for a bomb disposal man.

GERMAN STOCK MINE (CONCRETE)

Another German antipersonnel mine used is the **Stock mine**, meaning stick or picket mine. The mine is a cast-concrete shell containing pieces of shrapnel. The filling is a ⅓-pound charge of explosive.

The fuze is assembled by placing the explosive charge inside the casing and screwing the fuze and detonator into the top of the mine. This assembly is then placed on a picket projecting about 5 inches above the ground. Trip wires are fastened to the fuze.

The fuze for this mine can be either the ZZ42 or the ZZ35 pull fuze. When using the ZZ42 fuze, the trip wire is fastened to the actuating pin.

To disarm the mine, trace the trip wire to the mine and identify the fuze. If the ZZ42 fuze is used, carefully hold the actuating pin in the striker while another man cuts the trip wire. If the ZZ35 pull fuze is used, carefully insert a piece of stiff wire in the safety hole and then cut the wire.

FUZES—

WHAT THEY ARE, HOW THEY WORK

Fuzes are like the trigger on your gun; you cock them and they are ready to fire as soon as the safety is off and the trigger is pulled. Fuzes are convenient devices for setting off charges by any one or more of several ways. Mines usually use special fuzes designed for that particular type of mine. With standard types of fuzes, any kind of antitank mine, antipersonnel mine, or booby trap can be improvised. The enemy is only limited in his ingenuity by the materials at hand. We know a great deal about the German types of fuzes and how they are employed. The Japanese have not developed or designed many types of fuzes but have made great use of their antitank mines and grenades in booby-trap setups.

All the fuzes shown here need a detonator to set the explosive charge off. These are called nonelectric blasting caps. They fit onto the fuze and the charge has a well for inserting the detonator.

GERMAN PRESSURE FUZE D. Z. 35

This **pressure fuze** is made in two sizes, the larger shown here requiring 130–165 pounds to set off and the smaller size requiring only 65 pounds. The fuze is armed by removing the safety pin. To disarm, place a strong piece of wire in the safety-pin hole.

These four German fuzes are standard for antipersonnel mines, antitank mines, and for booby-trapping. The SMiZ 35 three-prong pressure fuze is normally used with the "S" mine; a pressure of about 15 pounds is required to set it off. This fuze, as are the following two types, is disarmed by replacing the safety pin or putting a stout piece of wire through the safety-pin hole. The ZZ35 pull fuze is made of brass and is used with "S" mines, to booby-trap antitank mines, and to fuze improvised explosive charges. A trip wire is fastened to the fuze striker end. The ZUZZ 35 combination pull and tension release fuze is similar in appearance to the pull fuze but is longer. If the safety pin is removed before the trip wire is taut and the safety-pin hole is centered in the oblong slot, the fuze will go off. The fuze is armed by fastening the trip wire to the fuze, drawing it tight until the safety pin is centered in the slot, and then carefully removing the safety pin. This is an extremely dangerous fuze and is seldom used. DO NOT TOUCH IT.

The ZZ42 fuze, unlike other fuzes, does not have a safety and will fire when the actuating pin is pushed or pulled out of the striker. This fuze is used in the Schu-mine, the Holzmine, and the Riegel mine. To disarm this fuze simply check that the actuating pin is in the correct position and then carefully remove the fuze from the mine or charge.

GERMAN TILT FUZE KI. Z. 43

The **tilt fuze 43 or** KiZ43 is the latest. It is designed to fire whenever the tilt rod is tilted in any direction. The fuze is intended for use on antitank mines, however, it is ideal for antipersonnel mines and booby traps. Only 1½ pounds pressure on the end of the extension rod sets it off. To disarm, replace safety using a nail or heavy wire; then unscrew fuze from charge.

DO NOT TOUCH TILT ROD.
GERMAN CLOCKWORK LONG-DELAY FUZE

This fuze has a clockwork assembly with a delay up to 21 days. The dials are inside the glass window. It has been used for delay charge left by the enemy around headquarters buildings, clocks, airfields, and power plants. The clockwork is started or stopped by turning the milled ring on its head so the red mark is at "steht" (stop) or "geht" (go). To disarm, turn the milled head so red mark is at "steht" and screw in plug in the side of stem; then unscrew the whole assembly from charge.

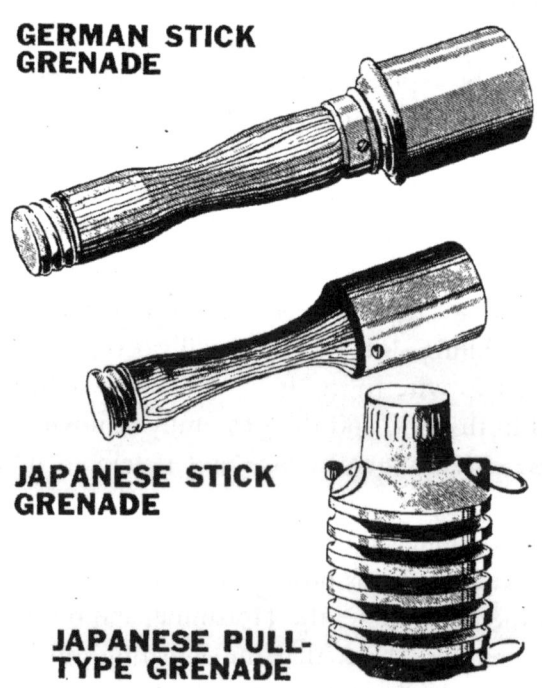

GERMAN STICK GRENADE

JAPANESE STICK GRENADE

JAPANESE PULL-TYPE GRENADE

These three grenades, the German stick grenade (potato masher), the Japanese stick grenade, and Japanese pull-type hand grenade, all work on the same principle. The stick grenade is armed by unscrewing the cap on the end of the handle and then giving a pull on the ring; this draws a friction wire through a match material, causing a flame that sets off a delay powder train of about 3 to 5 seconds, which in turn sets off the detonator and main charge.

The Japanese pull-type hand grenade is armed by depressing the cover catch, unscrewing the lead cover (1½ turns), and pulling on the firing string, which pulls the friction igniter through a match composition. Flame from the match composition ignites a 5½-second delay powder train, setting off the explosive. It is possible to remove the delay powder trains from these grenades so they will detonate instantaneously when the pull string is pulled. To dispose of these grenades, carefully search for booby traps near the grenade and then carry to a safe place and destroy by explosives.

The grenades described above are easily adapted as antipersonnel mines and booby traps. On the opposite page is a typical setup for this type of grenade as an antipersonnel mine. It is invariably used with a trip wire and is well camouflaged. It is set up by fastening the grenade to some solid object such as a tree or stake driven in the ground and tying the trip wire to the pull string in the grenade handle. The Japanese stick grenade is not readily converted into a booby trap by removal of delay train, as is the German stick grenade. The tar seal is difficult to remove and leaves obvious signs of tampering. To disarm, cut pull cord as short as possible without pulling it and place tape securely over opening.

JAPANESE 91 GRENADE
JAPANESE 89 HIGH EXPLOSIVE SHELL
JAPANESE 97 GRENADE

The Japanese grenades 97 and 91 are the same, except that the 97 grenade has a perforated propellent container which screws into the base allowing the grenade to be fired from the Japanese 50-mm grenade discharger (knee mortar). As grenades they are armed the same way. The safety pin is removed and a sharp blow on the pressure cap drives the firing pin into the percussion cap, igniting a delay train which in turn sets off detonator and main charge. The delays are usually from 4 to 7 seconds, but reports indicate the delay action is erratic.

The model 89 (1929) 50-mm high-explosive shell is fired from a grenade discharger. The fuze is safe until the safety pin is pulled out. It is armed by set-back when shell is fired. If detents have been removed, a slight blow on the point will detonate the shell.

All of these grenades can be used for antipersonnel mines and booby traps.

The Japanese grenades 97, 91, and model 89 can be employed as antipersonnel mines in any number of ways, but a sharp blow on the pressure cap is always necessary before the grenade will fire. The normal method of employing the grenades is under a pressure board. The Japanese have devised many schemes for using these grenades with trip wires. For example, trip wire when pulled releases the grenade so it falls far enough to cause the striker to fire the percussion cap, setting off the grenade. The model 89 grenade must have the detents removed before it can be employed in this manner. The 97 and 91 grenades can have the delay powder train removed so the grenade will fire instantaneously when the pressure cap is given a sharp blow. To disarm the 97 and 91 grenades, carefully replace safety pin or stout piece of wire through the safety-pin hole in the pressure cap.

JAPANESE BOOBY TRAPS

There is no doubt that the Japanese have information on German booby traps. Captured documents also indicate that the Japanese have their own booby traps.

Many items of regular Japanese ordnance can be adapted as booby traps. The 70-mm barrage mortar shell contains seven parachute bombs projected by a time train and fixed powder charge after the shell leaves the mortar. These can be made effective booby traps for the curious or unwary soldier either as captured materiel or if found on the ground as "duds." Grenades can be used for booby-trapping. For instance, a pull-type grenade can be fastened to a dead soldier with the pull string fastened to some solid object. Moving the body will set the grenade off.

Look out for electrically detonated booby traps. Any vehicle searchlight, generator, light circuit, or other electrical gear can be rigged easily so the current will detonate an explosive charge.

CHAPTER FIVE
WHERE DO YOU FIND THEM?
Everywhere!

ANTITANK MINES
are found in mine fields. . .
in roads, and along shoulders.

ANTIPERSONNEL MINES
are found not only in antitank and antipersonnel minefields, but also—
in bivouac areas. . .
in wire entanglements. . .
in likely routes of advance. . .
in obstacles.

BOOBY TRAPS are found wherever the enemy has been. . .
in mine fields. . .
in equipment. . .
in supplies. . .
in buildings. . .
in obstacles. . .
In fact you'll find booby traps in **ANYTHING** the enemy thinks you'll touch!

BOOBY TRAPS
have the same pressure, pull, and release-type devices as anti-personnel mines, but all sorts of schemes are used to set them off. Here are the more common ways of setting booby traps.

PRESSURE-TYPE FUZE

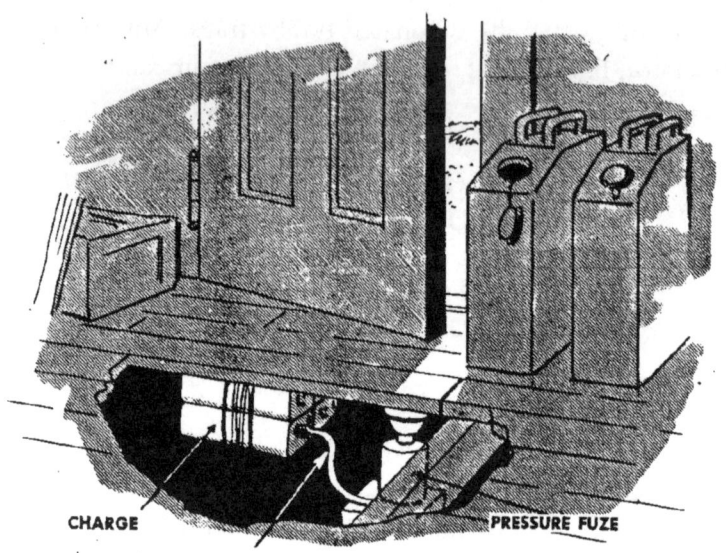

CHARGE

DETONATING CORD

PRESSURE FUZE

PULL-TYPE FUZE

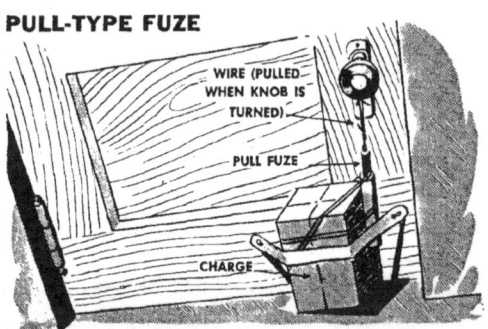

WIRE (PULLED WHEN KNOB IS TURNED)

PULL FUZE

CHARGE

RELEASE-TYPE FUZE

RELEASE FUZE

DETONATING CORD

CHARGE

ELECTRIC-CIRCUIT TYPE

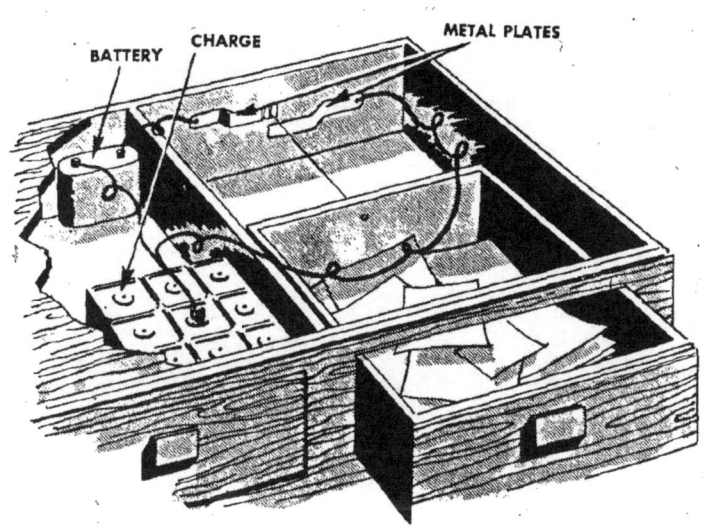

Wires or plates are brought into contact, completing an electric current which sets off an explosive charge. Batteries or current are necessary.

CHAPTER SIX
WHAT DO YOU DO ABOUT THEM?
You do and you don't!

What does the enemy want you to do?

He wants you to stop advancing.

He wants you to be confused.

He wants you to be afraid.

Don't play into his hands!

Believe all warnings. Stay in areas that are marked safe. Stay on roads and do not try to find a short cut. It doesn't pay.

If you have to go over ground that has not been cleared, carefully prod a path with your bayonet. Prod by pushing your bayonet into the ground at an angle. Do not jab; that might set off a Schü-mine. As you move forward, feel for trip wires. When you find any kind of a mine, try to find a way around. If you must remove the mine, get a 50-yard length of rope or signal cable, carefully tie onto the mine or trip wire, take a prone position at the far end of the line, warn all others in the vicinity to take cover, and pull out the mine.

Stay in marked lanes
Prod in unmarked areas!
Remove — with ropes!
[. . .]

What you DO–

1. Look where you're going.
2. Look at both ends of a wire before you touch it.
3. When you find a mine or booby trap, mark it, and report it to an officer or NCO.
4. Sandbag the driver's compartment of all vehicles.
5. Be especially careful at buildings and at road junctions, turnouts, parking areas, defiles, water points, and bypasses around road blocks and blown bridges.
6. Carry a 50-yard length of rope or signal cable in all vehicles.
7. Learn and observe [the following] marking signs:
Red triangle used on fences marking boundaries of mine fields.
Painted on both sides, used for marking safe lanes.
White mine markers placed over individual mines.

What you DON'T do –

1. Don't cut a taut wire; don't pull a slack one.
2. Don't attempt to disarm or remove a mine or booby trap unless you are trained to do so.
3. Don't move or touch abandoned vehicles, supplies, and equipment.
4. Don't drive or walk in areas not marked clear of mines.
5. Do not stand on running boards of vehicles.
6. Don't open doors or windows without first examining both sides.

Combat Engineers

While the term "combat engineer" can suggest something akin to a frontline infantryman with added engineering skills, the duties of the U.S. Army's combat engineer battalions were as characteristically broad as other elements of the military engineering profession. A combat battalion, the 1943 composition of which is described below in the publication FM 5-6, Operations of Engineer Field Units, *was a unit organic to each U.S. Army division. The "combat" adjective denoted the fact that the battalion was deployed in support of combat operations. At its most aggressive, this could mean engineers participating directly in assault actions—as demolition experts or flamethrower teams during attacks on enemy positions, for example. Indeed, there were many occasions during World War II when USACE troops were used, in extremis, in pure infantry roles, which they tended to do very convincingly. But often the combat engineers were required to provide support services to keep the division moving and operating, especially by opening road transport routes, repairing or laying bridges, clearing minefields, and creating important structures, such as artillery positions or supply depots.*

In the passage of text below, we see how seamlessly the duties of infantryman could be interweaved with those of the engineer. What the engineers brought to infantry combat, however, was an additional layer of tactical and technical insight, particularly in relation to skills such as making and finding cover, the employment and positioning of heavy weapons, the use of explosives, and how to overcome tactical obstacles with engineering solutions. But the work of the frontline combat engineer was often of necessity improvisational, oriented to speed rather than perfection: "Construction work done under combat conditions by division engineer units must necessarily be of a hasty and temporary nature, fitting the minimum

Corporal Robert Lopez of Company D, 532nd Engineer Amphibious Regiment, prepares to embark for Europe in 1943. (Signal Corps Archive)

requirements of the immediate tactical situation. When the situation permits, such construction is usually supplemented or replaced by work of a more adequate and permanent nature executed by general engineer units under control of the unit engineer of a higher echelon" (FM 5-6, p. 259). For the combat engineers, the crucial objective was to get the job done, by whatever means were available.

★★★

From FM 5-6, *Engineer Field Manual: Operations of Engineer Field Units* (1943)

CHAPTER 4
COMBAT OPERATIONS

SECTION 1
EMPLOYMENT OF ENGINEER UNITS
FOR DISMOUNTED COMBAT

■ 35. GENERAL.—*a. Combat engineer troops.*—(1) In carrying out their mission, engineers may become involved in combat. They may participate actively in defense against airborne and mechanized troops, in the defense of road blocks and mine fields, in tank hunting, and assist by demolition in the passage of hostile obstacles and the capture of fortified localities. Often they are forced to fight to maintain their own security while on the march, in bivouac, or at work. In combat of the nature outlined above, engineers may fight in small units—a squad or less, a platoon, or occasionally, a company. They should be trained thoroughly in the use of their weapons and in infantry tactics.

(2) In an emergency, engineers with a division may be relieved of their engineer work and used for combat. In such case the decision must be weighed carefully by the division commander, as the cessation of engineer work may result in a loss of mobility for the division and thus a reduction in its total combat power not compensated for by the use of engineers as infantry.

(3) This section describes the organization and employment of engineer units for combat in the execution of missions normally performed by infantry. Succeeding sections describe combat operations of engineers in providing for the security of their own and larger units, in defense of obstacles and barriers, in passage of artificial obstacles, and in assault of fortified localities.

b. Aviation engineer troops.—Aviation engineer troops are equipped and trained for combat primarily so that they may assist in the defense of airdromes. They may be used to defend the airdrome itself against hostile airplanes, parachutists, airborne troops, and ground raids, or to attack any force that may have established itself nearby.

c. Other engineer troops.—Other general and special engineer troops should be trained to provide their own local security when at work, in movement, or in bivouac, and to assist in the defense of rear installations when they are subjected to sudden airborne or mechanized attack.

d. Basic tactics.—The basic tactics of engineers when engaged in combat are the same as those of infantry.

■ 36. GENERAL ORGANIZATION FOR COMBAT.—*a. Modification of normal organization.*—When employed in combat either in furnishing their own security during engineer missions or in action on missions normally assigned to infantry, engineers do not reorganize to conform to the organization of infantry but enter combat with their normal organization. Minor modifications are made as necessary to provide for the effective employment of the crew-served weapons, for the security of engineer equipment not needed for combat, and for the special problems of command, communication, and ammunition supply incident to combat. A standing operating procedure should be established by all units, including platoons, fixing definitely such modifications as are considered essential.

b. Division into echelons.—When an engineer unit enters combat as a whole it is divided into a forward and rear echelon. The forward echelon comprises those elements that actually engage in combat and also the command, communication, and ammunition supply personnel and equipment necessary to control and supply the combat elements. The rear echelon consists of the personnel and equipment not needed for combat. It includes kitchen and water trucks; trucks carrying engineer supplies and equipment; and special vehicles such as air compressors and tractors. The minimum personnel necessary to maintain the mobility of the rear echelon, provide for its local security, and perform essential administrative functions is assigned.

(2) The actual composition and location of the rear echelon varies with the situation and the size of the unit. Small engineer units operating

alone often have it close at hand with only the vehicle drivers for protection. Larger units, such as an engineer battalion participating in the attack or defense of a position, usually have the rear echelon a considerable distance to the rear. The personnel required for its protection varies but should be kept to a minimum. In most situations, the light vehicles are required with the forward echelon for purposes of security, communication, ammunition supply, and the displacement of heavy weapons.

c. Continuation of engineer operations.—The engineer operations of the unit normally are suspended when the unit is committed to combat. However, certain types of engineer work, such as the operation of water points and engineer reconnaissance, may have to be continued by personnel of the rear echelon.

A soldier of the 112th Engineers conducts air defense drill with an M1917 Browning machine gun. (Signal Corps Archive)

■ 37. ORGANIZATION OF COMBAT BATTALION FOR COMBAT.—*a. Combat platoon.*—(1) When organized for combat, the platoon consists of a platoon headquarters; a weapons section to operate the caliber .30 machine guns and the caliber .50 machine gun; and three rifle squads, each including rocket launcher. Combat duties to maintain communication and control are assigned to some members of the platoon headquarters and crews are provided for the heavy weapons, these crews being drawn from the remaining members of platoon headquarters and the three squads of the platoon.

(2) The men provided in the T/O [Tables of Organization and Equipment] whose sole duties are the operation and control of the platoon heavy weapons are insufficient to provide crews for all of them; therefore, it is necessary to assign men from personnel normally employed on engineer duties. Each combat engineer squad should have a corporal and three men trained as a crew to operate a machine gun. This permits a machine gun to be assigned to any squad for operation and also makes available trained men who can be drawn from the squads to provide crews when the weapons are operated under platoon, company, or battalion control. Each machine-gun crew should be trained to operate either the caliber .30 or the caliber .50 machine gun or any type of direct-fire mission.

(3) The size of the crew provided for a machine gun varies according to circumstances. A machine gun fired from a vehicular mount may be served adequately by two men. In static situations or where only limited movement by hand is required, the efficient operation of either of the machine guns on its ground mount requires four men; and in situations where the weapon must be moved by hand frequently and over considerable distances, a crew of six men normally is required. When the situation requires a large crew, consideration should be given to employing a reduced number of weapons in order not to deplete the rifle squads unduly.

b. Combat company.—The forward echelon of a combat company employed in combat consists of three platoons, the communication and supply personnel necessary to assist the company commander, and often, depending on the method of control and operation of the machine guns, a provisional heavy weapons platoon. The rear echelon usually is commanded by the company motor sergeant.

c. Combat battalion.—For combat, the battalion is organized into a forward and a rear echelon. The major portion of the headquarters and service company is assigned to the rear echelon. The forward echelon of the headquarters and service company consists of the personnel necessary to assist the battalion commander and his staff in the combat duties of command, intelligence, operations, signal communication, and supply. The commander of the headquarters and service company normally remains with the rear echelon of the battalion, and the rear echelon of each of the lettered companies usually are placed under him. When some of the machine guns of the battalion are withdrawn from the companies to operate directly under the control of the battalion commander, it is often necessary to provide for their command and control by personnel of the forward echelon of the headquarters and service company.

■ 38. Employment of Engineers in Offensive Combat.—*a. Fire and movement.*—Foot elements attack by a combination of fire and movement. Aimed fire destroys or neutralizes the enemy and protects movement on the battlefield. In engineer units the fire is provided by machine guns and rifles. Rifle elements move by infiltration, by following covered routes, or by alternate rushes of individuals or small units, in order to establish themselves at closer ranges where more effective fire can be delivered. Enveloping action and positions from which converging and flanking fires can be delivered are sought by preference, even by small units. Rifle elements combine their fires with those of the machine guns, and exploit by alternate fire and movement the resulting neutralization of the enemy. The aim is to close with the enemy and destroy him with the bayonet and other close-range weapons. Units most favored by terrain or supporting fires push forward, while those most exposed support the advancing elements by fire.

b. Technique of fire.—Supporting weapons are kept well forward where the crews can see their targets and can see from the close vicinity of the weapons emplacements the location of the units supported. Fire of all elements is observed fire against point or linear targets. There is not sufficient ammunition and materiel available to permit covering extensive area targets.

c. Security.—Security measures are continuous. Contact is maintained with adjacent units. Detachments are provided to observe and protect exposed flanks and the rear. Antitank weapons are posted to cover approaches favorable for hostile tank attack, their dispositions being coordinated for the entire unit. Antiaircraft and antimechanized warning service is maintained. The principal antiaircraft weapon in offensive action is the rifle. Machine guns not required for essential ground tasks should be mounted for antiaircraft protection of important elements of the unit.

d. Formulation of plans.—The commander prescribes a detailed tactical plan only for that portion of the operation in which he can estimate the hostile resistance. A small unit cannot plan as far ahead as a larger unit because the situation confronting it changes more rapidly. For units smaller than a battalion, it is sufficient to prescribe initially only a simple combination of fire and movement for the capture of ground which can be seen. Simplicity is the keynote of orders and actions. Simple, definite tasks are assigned to the various elements to coordinate their action. Surprise is an essential element of a successful plan of attack. It is gained by concealing the time and place of attack, screening the dispositions, employing rapidity of maneuver, and avoiding stereotyped procedures. The plan of attack should provide for a reserve to be held out initially, and used later to exploit weakness discovered in the hostile position or to take over the assault role if the assaulting echelons are stopped. The reserve, when used, is preferably employed for enveloping action.

e. Employment of combat battalion.—(1) *General.*—(*a*). When engineer units are used as infantry in the attack of a position, normally the smallest unit so employed is the combat battalion. It usually operates under the direct control of the division commander and is assigned a zone of action and an objective by him. Supporting fires beyond the capabilities of the organic weapons of the battalion are normally required. Artillery support is usually furnished and in some cases combat aviation and tanks.

(*b*) The battalion commander decides on a plan of maneuver, arranges for supporting fire, and coordinates the action of the companies and any supporting weapons under his control by assigning specific missions to them for each phase of the attack. He assigns objectives and zones of action to each company in the assault echelon. The objectives assigned the companies may be parts of the battalion objective or of an intermediate objective selected by the battalion commander. When such an intermediate objective is captured the supporting weapons are brought forward to establish a new base of fire and the battalion is hastily reorganized to attack the next objective in accordance with a new plan prescribed by the battalion commander.

(2) *Employment of antimechanized weapons.*—Antimechanized protection for the battalion is provided primarily by rocket launchers. From time to time during the progress of the attack the battalion commander assigns specific responsibilities to the companies in this connection. He does this by designating the particular directions or approaches for which each company is responsible. It is unusual to concentrate a part or all of the rocket launchers under battalion control for this purpose. The antitank grenade is primarily for the local protection of the units armed with that weapon. The caliber .50 machine gun has a limited effect against tanks and requires comparatively large crews which in an attack can usually be more effectively employed elsewhere. Because of this, the employment of the caliber .50 machine guns for antimechanized defense during the attack rarely is desirable. When so employed they normally are organized into improvised platoons and sections and handled under battalion control.

(3) *Employment of caliber .30 machine guns.*—In offensive action by units larger than the platoon, effective use of the caliber .30 machine guns normally requires control by company and at times in part by battalion. Usually most of these weapons are controlled directly by the company commanders. Some guns, however, may be withdrawn from the companies and operate directly under the battalion commander or under a temporary commander especially designated by him. Operation of all machine guns by the battalion in this manner is rare, since improvising an organization to control so many weapons is impracticable. Six guns is usually the maximum number used in this manner. Centralized control of machine guns in offensive action is most appropriate in open,

Engineer Herschew Crafton of the 10th Engineers, 3rd Division, seen shortly before embarkation with Task Force A for North Africa, October 1942. (Signal Corps Archive)

rolling terrain where conditions are favorable for coordinating and massing supporting fires. In close, broken terrain more decentralization is required, and the guns are operated by companies and even platoons. Frequently, in order to add mobility to the attack, only eight or ten of the caliber .30 machine guns are employed, the crews which would have served the others being used as riflemen.

[. . .]

■ 42. EMPLOYMENT OF ENGINEERS IN DEFENSIVE COMBAT.—*a. General dispositions.*—Use of engineers in place of infantry for the defense of a position is an emergency measure, continued only for short periods. When so used, engineers operate directly under the division commander. Because of differences in strength and armament, a combat engineer battalion employed as infantry in defense of the main line of resistance should not be assigned a frontage greater than two-thirds that assigned to an infantry battalion under similar conditions.

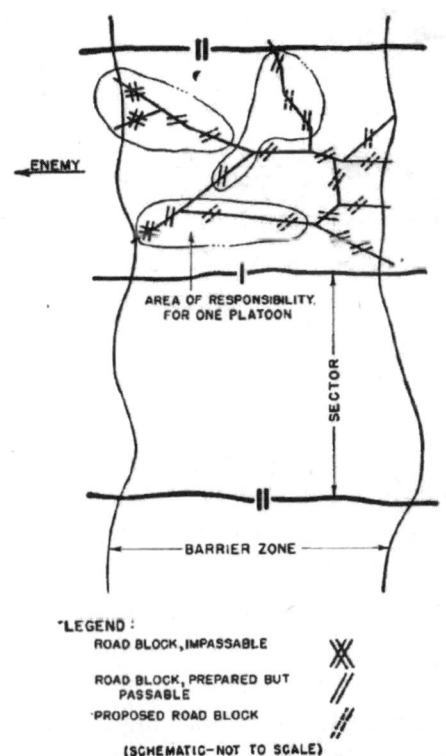

FIGURE 4.—Initial dispositions of an engineer company for defense of a sector of a barrier zone by delaying at successive obstacles.

b. Conduct of the defense.—The defensive mission of a unit on the main line of resistance is to stop the enemy by fire in front of the battle position, to repel his assault by close combat if he reaches it, and to eject him by counterattack if he succeeds in entering it.

(1) Surprise is sought. Every effort is made to keep the enemy in doubt as to the location of the main line of resistance and the principal elements of the defense. Camouflage, varying the defensive dispositions, having alternate positions for small units, and screening action by combat outposts contribute to this end.

(2) The general trace of the main line of resistance and the lateral limits of the battalion area are designated by higher authority. The battalion is disposed to take advantage of observation, fields of fire, concealment and cover, natural obstacles, and routes of communication.

(3) Company defense areas are designated by the battalion commander. Small defended localities, the smallest of which normally is garrisoned by a platoon, are located and organized so as to be mutually supporting throughout the width and depth of the battalion area. In each defended locality and for the battalion as a whole provisions are made for all-around defense.

(4) The small defended localities are located to cover strongly the important approaches to and into the battalion position, to cover gaps in the plan of machine-gun and supporting artillery fires, to protect the supporting weapons in the area, and to provide mutual support.

(5) Elements on the main line of resistance normally fight in place with only those minor changes in position necessary to meet envelopments. Company supports and battalion reserve elements are primarily holding forces, but are employed, when necessary, for local counterattacks. Possible counterattacks must be foreseen and plans made for them.

(6) The battalion machine guns are disposed according to a coordinated scheme for the battalion as a whole. Alternate positions are necessary.

(7) Regardless of any covering forces or outposts provided by higher authority, local security detachments are posted at short distances to the front and on exposed flanks to give warning of enemy approach.

(8) The battalion depends, in great measure, for its antiaircraft protection on the concealment and cover afforded by natural features or

entrenchments, on the dispersion of its elements, and on the antiaircraft fires of the automatic weapons of its supports and reserve.

c. Antitank weapons.—The employment of antitank weapons is coordinated by the battalion commander. He does not normally improvise a command organization for them as actual command and control normally are best exercised by the company and platoon commanders. Normally he assigns positions and sectors of fire for such caliber .50 machine guns as may be employed and designates the directions or approaches which each company is responsible for covering with the rocket launchers and antitank grenades. In order to make a more effective distribution of antitank weapons to fit the terrain, it may be desirable to take caliber .50 machine guns and rocket launchers from one company and assign them to another. Particular attention is paid to the protection of units on the main line of resistance.

d. Caliber .30 machine guns.—In defense, the coordinated employment of caliber .30 machine guns by the battalion is habitual and the method of control described in *c* above for the antitank weapons is used. The machine guns are sited so as to cover the front, flanks, and depth of the battalion area with mutually supporting fires. Machine guns in the forward part of the area are sited primarily for flanking fires in close support of elements on the main line of resistance. Normally they are sited in pairs. Units smaller than a battalion acting alone in the defense also employ unit control and coordination of machine guns. In smaller units it may be necessary to site guns singly.

[. . .]

Section IV
PASSAGE OF ARTIFICIAL OBSTACLES

■ 53. General.—*a.* This section deals with the passage of artificial obstacles. Common types are barbed-wire entanglements; mine fields; prefabricated obstacles such as tetrahedrons and concrete blocks; steel, wood, and concrete post obstacles; timber obstacles such as abatis and cribs; and antitank ditches. [. . .]

b. In the passage of obstacles, the first step is usually to neutralize the defending fires. Generally other arms do this and then engineers or other specially trained troops are employed to clear the obstacle. If resistance is overcome in advance, the engineer task is greatly simplified. However, in an attempt to gain surprise, preparations for the breach may be made in secrecy or under cover of darkness, with or without supporting action. In many cases engineer work must proceed simultaneously with the combat action of other arms. This necessitates maximum coordination which generally requires a rehearsal by all troops taking part.

■ 54. METHODS OF CLEARING OBSTACLES.—The methods of forcing a passage through an obstacle may be grouped under five main headings: destruction by hand-placed explosive charge, hand removal, destruction by artillery fire or aerial bombardment, surmounting or bridging, and destruction by fire of tank weapons.

a. Destruction by hand-placed charge.—Breaching an obstacle with hand-placed charges is an efficient method. Obstacles of concrete, wood, or steel are readily destroyed by using explosives in accordance with standard demolition practices. Hand-placed charges are also effective for clearing gaps in mine fields. After the individual mines are located by mine detectors or by probing and hand searching, small charges are placed on each mine and connected by means of primacord to a main line of primacord running along the center line of the proposed gap. All mines in the gap are then destroyed when the main line of primacord is detonated. The firing of charges that have been placed in secrecy or under cover of darkness should be timed so that the breaches in the obstacles will be used immediately. Bangalore torpedoes are used extensively in breaching barbed-wire entanglements, and may also be used for clearing gaps through mine fields.

b. Hand removal.—The clearing of gaps in minefields for the passage of tanks can be accomplished by hand removal. In this method the mines are first located by probing or by mine detector. If tanks are to lead the attack this is done secretly the night before. If the infantry leads, as is normal, it is done as soon as an infantry bridgehead is established. After being located and marked by trained specialists, the mines are disarmed and removed. Experienced men carefully uncover the mines, ordinarily by digging with their hands, determine the type of mine and nature of any auxiliary firing device, disarm the mines, and remove them. Antitank ditches can be quickly and effectively prepared for passage by pulling down the vertical banks with picks and shovels. The hand removal of well-constructed concrete, wood, or steel obstacles usually requires too much time and cannot be done secretly.

c. *Destruction by artillery fire or aerial bombardment.*—Reliable results in breaching obstacles cannot be expected from artillery bombardment. Some obstacles may be deranged and some destroyed by direct fire at short ranges. If indirect fire or bombs are used, the expenditure of ammunition will be large in proportion to results obtained. However, artillery preparatory fires often will discover the location of mine fields by exploding some of the mines.

d. *Surmounting or bridging.*—Surmounting or bridging is impracticable for crossing enemy minefields, but is effective in passing certain forms of tank or personnel obstacles, particularly ditches and craters. Wire netting or mats may be used for surmounting barbed-wire entanglements. Heavy crater charges placed in roads by the enemy may be bridged unless it is quicker to remove the charge, or to explode it and then repair the road.

e. *Bombardment by tank weapons.*—Obstacles which are susceptible to breaching by light artillery fire may be attacked directly by the fire of medium tanks.

f. *Method of breaching.*—The decision as to the method of breaching the obstacle is made by the commander of the attacking force after consultation with his engineer. The decision is based upon the mission; type of obstacle; time required; requirements of secrecy; type and strength of defending fires; and materiel and support available.

[. . .]

■ 57. EMPLOYMENT OF ENGINEERS.—*a. Assignment.*—Engineer units clearing obstacles may operate under engineer control or may be attached to another unit. Where the operations are for the purpose of facilitating the movement of a single unit, such as an infantry regiment or tank battalion, the necessary engineers are attached to that unit and revert to their own command after the tasks for which they were attached are

Men of the 1139th Engineer Combat Group take a moment of rest between operations while bivouacked in France, August 1944. (1139th Engineer Combat Group)

accomplished. However, the commander of the engineer unit from which the smaller units are detached should supervise carefully to insure that the detached engineer elements carry with them the supplies needed.

b. Organization of clearing parties.—(1) The strength, organization, equipment, and means of transportation for each clearing party must be carefully planned considering its mission and the known characteristics of the obstacles to be cleared. Parties may be organized to clear obstacles in front of the enemy main line of resistance at the beginning of the attack; to accompany tanks or front-line infantry units; to clear obstacles encountered within the enemy position after the attack begins; or to mark and remove obstacles already passed by the leading elements. Clearing parties must have the same mobility as the units they accompany. Clearing parties accompanying infantry generally move on foot and should not carry such heavy loads of engineer equipment and supplies that they will not be able to keep up. Engineers accompanying tanks normally are transported in trucks or tanks. Supplies carried must be carefully planned so as to provide all essentials and eliminate nonessentials.

(2) Clearing parties normally are organized into task groups. When the obstacle to be cleared is formidable and complete information about it is available, the organization for clearing the obstacle can be highly specialized and the men forming each specific task group can be instructed in detail and rehearsed in the performance of precise duties. However, to cope with obstacles which will be encountered after the attack begins and concerning which only incomplete information will be available, a flexible organization must be adopted. Groups to perform each of the following functions may be included in a party; clearing antipersonnel mines, breaching the obstacle, marking the gap, providing local security, laying smoke locally for protection of clearing operations, and providing replacement or reinforcement of other groups.

c. Conduct of clearing operations.—Passages cleared through obstacles must be located at exactly the places ordered by the higher commander. This is the responsibility of the commander of the clearing party. A marking detail generally is organized for this, utilizing issue marking equipment or improvisations. A breaching detail may be preceded by a group to clear away antipersonnel mines. A breaching detail may employ hand-placed charges, hand removal, bridging or surmounting for the passage of primary obstacles. The mines are removed by hand or destroyed in place by explosive charges or mechanical means. When the obstacle or obstacles have been breached, the marking details mark the boundaries of the gaps created and the routes between gaps in successive obstacles. When enemy interference with the clearing operations is likely, the support group provides local security. In such a case, the clearing party may lay down a smoke screen, but normally smoke required is laid by other troops. Reserves of men and materials are provided to replace losses.

■ 58. Rear-area Obstacles.—Engineers accompanying troops that have broken through, enveloped, or been dropped in an enemy position are used to overcome rear-area obstacles. Most of these obstacles will not be so thoroughly covered by enemy fire as front line obstacles, and the engineers will be able to clear passages more rapidly. In order to clear obstacles with the least delay, demolitions are extensively used.

[. . .]

CHAPTER 5
RIVER-CROSSING OPERATIONS

Section I
TACTICAL PROCEDURE

■ 68. GENERAL.—*a.* The immediate object of the attack of a river line is to establish one or more bridgeheads to protect the crossing of the remainder of the command. A division usually crosses as part of a larger force, with one of the following missions: to force the main crossing, to make a secondary crossing, or to feint.

b. One or the other of two general situations normally confronts a division planning to cross a stream over which the enemy has destroyed the bridges. In the first, the opposite bank of the stream is strongly held by the enemy and a deliberate river crossing is necessary. In this situation the crossing must be made on a broad front and powerful fire support must be organized. In the second, the opposite bank is lightly held by the enemy. When this is the case, the division should cross a portion of its strength immediately, to establish and hold a bridgehead to cover the crossing of the remainder of the division. Such an operation is called a hasty river crossing.

c. Reconnaissance, planning, and assembly of necessary forces take place during the preparatory phase of the operation. Thereafter, the crossing of each element is conducted in four steps: movement forward from initial assembly areas to final assembly areas; movement from final assembly areas to the crossing points; crossing; and attack on far bank to seize successive objectives.

d. The force landed on the far bank usually has three successive objectives:

(1) *First.*—A position the capture of which will eliminate effective direct small-arms fire upon the crossing front.

(2) *Second.*—A position the capture of which will protect the selected ponton-bridge site from ground-observed artillery fire and which can be supported by light artillery on near side of the river.

(3) *Third.*—A position the capture of which will protect the bridge site from all artillery fire and will provide proper maneuver space on the enemy side of the river.

[. . .]

■ 75. CROSSING BY ASSAULT BOATS.—*a. Procedure.*—Under cover of darkness, all troops go into their selected positions. The assault boats are brought to the final assembly areas by truck when the terrain, the road net, and the requirements of secrecy permit. Engineer troops unload the boats, carry them to the boat-group assembly areas, and distribute them along the foot routes to the river so that they may be readily picked up by the infantry carrying parties. They place the boats from 100 to 200 yards of the river edge unless lack of cover requires a greater distance.

(1) Engineer boat groups are organized [. . .] each group consisting of a reinforced engineer squad with four or five assault boats.

(2) Leading infantry units move to the final assembly area where they are met by the engineers who lead boat groups to the boat-group assembly areas. At the boat-group assembly area, each infantry boatload is guided to its boat. The infantry troops assigned to a boat carry it down to the river bank, led by the engineer guide. Upon arrival at the embarkation points, the troops with each boat embark immediately and paddle to the far bank. There they get out, overcome any obstacles found, and neutralize enemy resistance near the bank. Special equipment, such as chicken wire and bangalore torpedoes, may be carried in assault boats by leading elements to facilitate passage of obstacles not eliminated by fire. After neutralizing enemy resistance near the bank, the leading infantry units advance to the first objective.

(3) Departures from the final assembly areas are timed to permit leading units to cross simultaneously on a broad front, but once these units leave

the final assembly areas they do not halt for coordination and no attempt is made to maintain alinement between boat groups. In night crossings, firing from boats is prohibited.

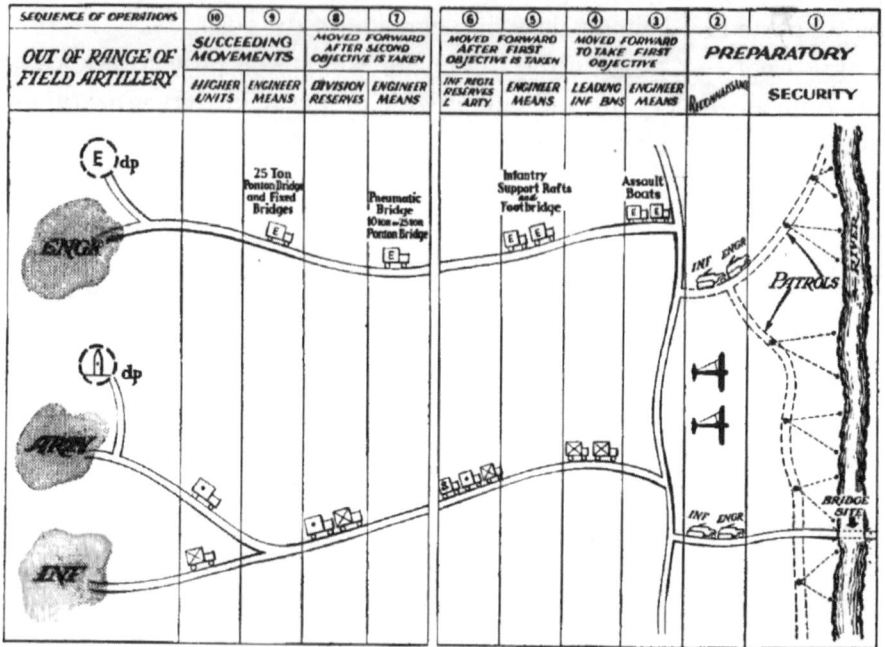

FIGURE 8.—Sequence of operations, river crossing.

b. *Handling assault boat.*—(1) Ten to twelve men carry the boat. From the time of departure from the final assembly area until the moment the far bank is reached, no unnecessary stops are made. The boat is carried forward inverted, as far as a spot previously selected by the engineer leader within a few yards of the water edge. At this spot the boat is righted. Great care must be exercised to insure that silence is preserved. Objects must not be allowed to strike the sides or bottom of the boat. Rifles should be slung diagonally, muzzle up, from the shoulder nearest the boat. Special care must be exercised in carrying the boat with its bottom down and every precaution must be taken to prevent it from striking stumps, rocks, and other obstructions and from being dropped or dragged on the ground.

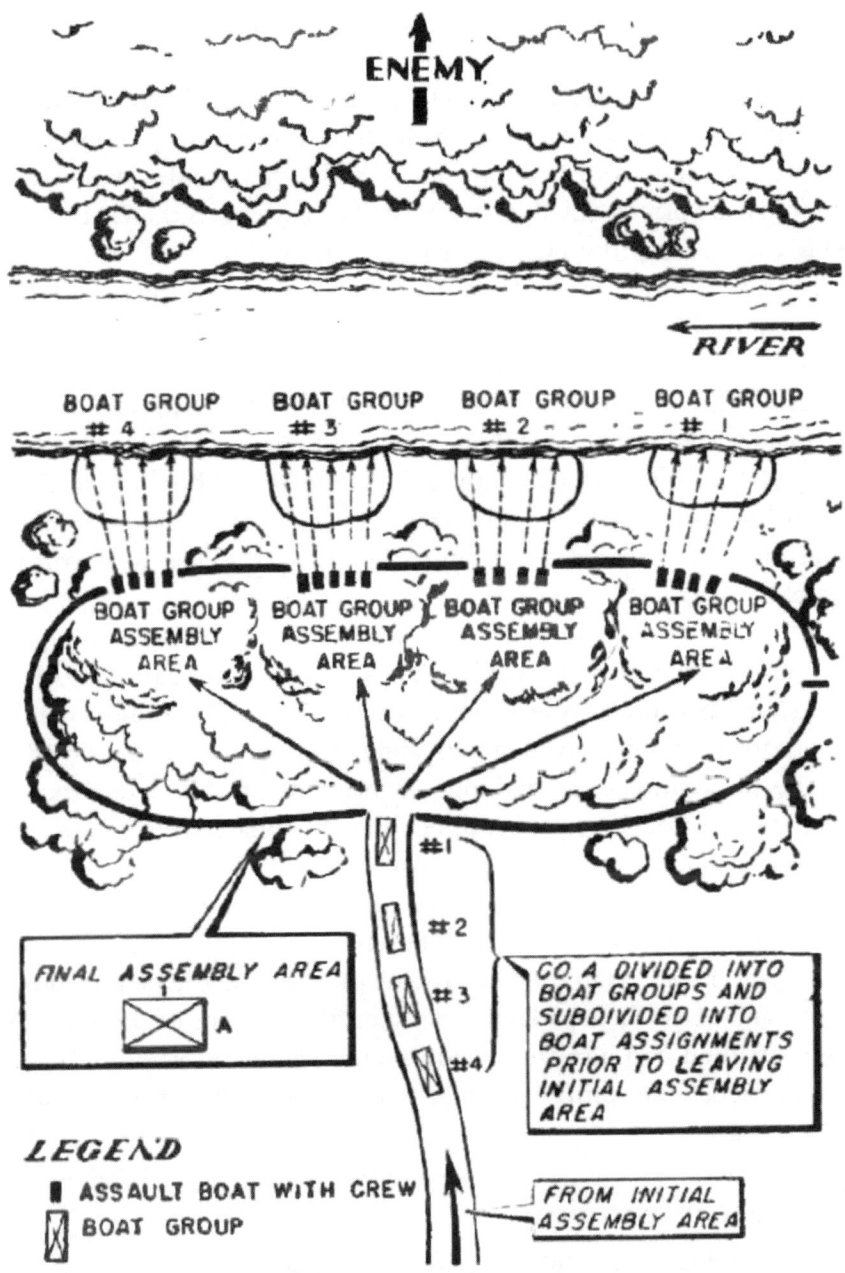

FIGURE 9.—Final assembly area.

(2) Immediately upon arrival at the river bank, and without change in the carrying formation, the boat is carried bow first into the water until a depth of water sufficient to float the fully loaded boat is reached. The boat is loaded parallel to the bank if the water is deep immediately offshore. Ammunition, machine guns, and other similar weapons or equipment are placed quietly in the boat. The passengers then get aboard, taking care to keep the boat in balance and avoid noise. Two engineer soldiers take position in the bow, one on each side, and the third kneels in the stern. The soldier in the stern is in charge of the boat and does the steering. The infantrymen at the sides paddle the boat, the engineer paddlers in the bow acting as strokes. Paddlers, whose rifles are held by the infantry soldiers not paddling, kneel on their outside knees and use their paddles without touching the sides of the boat. Those not paddling crouch low in the boat, holding rifles upright against the bottom. All men should be prepared to slip off packs in case the boat is overturned or sunk.

(3) Each boat starts across as soon as loaded and is paddled as rapidly as possible to the opposite bank. Unless specific orders to do so have been issued because of the relative positions of available embarkation and debarkation points or the width and current of the stream, no effort is made to counteract drift. Under conditions of complete darkness, heavy fog, or smoke, the proper direction of the boat may be maintained by use of a luminous compass. Upon arrival at the far shore, silence is preserved unless the enemy has discovered the crossing and opened fire. To avoid noise, the boat is not beached, except on a mud bottom. The engineer soldiers in the bow disembark and hold the boat. Paddlers get out and place their paddles quietly on the bottom of the boat. All passengers then step into the shallow water or directly ashore. Cargo is then unloaded. The entire crossing is executed with the utmost rapidity.

Combat Integration

The following is part of an after-action report from the HQ, 28th Infantry Division, which notes the integral role played by combat engineers in divisional operations.

(a) In the southern part of the Division sector, 109th Infantry, with Company A, 103d Engineer Combat Battalion, Company C, 707th Tank Battalion, and 107th Field Artillery Battalion attached, met and contained all enemy penetrations. TANDEL (P8845) was retaken but LONGSDORF remained in enemy hands.

(b) The enemy continued to infiltrate in the northern part of the sector and a task force made up of a platoon of Infantry, a Platoon of Tanks, and a Platoon of Engineers was sent to block the roads leading South from HOSCHEID.

(c) In the central sector the 110th Infantry, with Company B, 103d Engineer Combat Battalion, Companies A and B, 707th Tank Battalion, 1st Reconnaissance Platoon, 630th Tank Destroyer Battalion, and 109th Field Artillery Battalion attached, was attacked throughout the day by strong enemy tank and infantry forces. The 1st Battalion held strong points in the sector, but was unable to stop enemy penetrations into CLERF (P7763). CLERF was cleared of enemy at 1130A, but by 1800A twenty-five (25) enemy tanks followed by infantry penetrated into CLERF, surrounded the Regimental CP, and brought direct tank fire to bear. Communications were lost and the situation with reference to Regimental Headquarters personnel was obscure. The 2d Battalion (- Company G) counterattacked from the high ground Northeast of CLERF towards MARNACH, but their situation was obscure due to encirclement by the enemy. Company G reverted to Regimental control at 1640A and was moved to the vicinity of ESELBORN (P7764). Elements of the 3d Battalion, 110th Infantry were organized into a strong point at CONSTHUM (P7954) by the Regimental Executive Officer [Lt. Col. DANIEL B. STRICKLER]. Contact with Company K, which was at HOSINGEN (P8259) was lost at about 1600A and at dark it was believed that the enemy held the town.

(d) Early the morning of 17 Dec 44, Company D, 707th Tank Battalion attacked South from WEISWAMPACH along the ridge road toward MARNACH to relieve the situation there. They fought their way South against enemy numerical superiority in tanks and infantry and only after losing fifteen (15) light tanks did they withdraw.

(e) In the northern sector the 112th Infantry with Company C, 103d Engineer Combat Battalion and 229th Field Artillery Battalion attached, withstood heavy enemy attacks throughout the day, but several penetrations had been made and under cover of darkness the 112th Combat Team withdrew to the West Bank of the OUR River and took up a defense of the river line in assigned sector.

(f) Late in the evening of 17 Dec 44, the enemy penetrated to within three kilometers of the Division CP at WILTZ and the headquarters of the

707th Tank Battalion (P7656) was overrun by the enemy and they fought a delaying action towards WILTZ.

(g) The 44th Engineer Combat Battalion was given the mission of the defense of WILTZ and took up a defensive position around the city at 1800.

★★★

The Engineer Soldier's Handbook *contains several spirited sections that explain how engineers could apply themselves to destroying hardened enemy targets, including armor and fortified positions. While the tactical principles described are sound, we should take them with a pinch of salt at times, with the knowledge that an intelligent enemy had its own countermeasures against such threats. For example, German armor would rarely operate without at least some protective infantry support to deal with close-range attacks by enemy engineers or infantry. Yet it is undeniable that the U.S. Army engineers could be highly motivated and effective foes against Axis forces. During the German Ardennes Offensive of 1944–45, for example, U.S. engineer units opposed German armored columns through a combination of blowing bridges, creating road obstacles, laying hasty minefields, making direct attacks with bazookas and antitank guns, and generally imposing delays on forward movement, which in turn helped to consume the Panzers' rapidly diminishing fuel supplies. Engineer ingenuity, applied in close cooperation with other combat units, could make a very challenging environment for enemy operations.*

★★★

From FM 21–105, *Basic Field Manual: Engineer Soldier's Handbook* (1943)

CHAPTER 7
ENGINEERS AND TANKS

SECTION I
TANK HUNTING

■ 56. THE ADVANTAGES ARE WITH THE HUNTER.—The big game sometimes hunted by engineers are tanks. Like any other kind of big game hunting, such as elephant hunting or lion hunting, the advantages are with the hunter; he almost always is the winner; but there is enough danger in the sport to keep the hunter on his toes. With courage and determination the engineer can use his weapons to hunt down and destroy 80,000 pounds of fighting steel.

■ 57. TANK WEAKNESSES.—Tanks have a number of weak points and any one of these may be used for its annihilation. Here are some things to remember:

a. A tank is big—a large target.

b. A tank is run by a mechanism which is breakable.

c. A tank is armored, but there is a limit to its armor and our weapons are capable of piercing the heaviest armor.

d. Tanks can't go everywhere. They can't climb steep banks, hurdle special obstacles, ford deep streams, or go through thick forests.

e. Tanks are partly blind. They can't see as well as you can.

f. They can't go over a mine undamaged.

g. They are run by human beings—men as vulnerable to fire, lead, steel, heat, and explosives as a man out of a tank.

h. Tanks are noisy; they can't "sneak up" on you, and they can't hear most noises.

■ 58. ANTITANK WEAPONS.—The weapons of the engineer tank hunter are simple to use, but deadly in their effect. Some of the weapons used are—

a. Antitank gun, 37-mm.—The 37-mm gun is a high-velocity weapon. Its armor-piercing ammunition penetrates all but the largest tanks. It is extremely accurate and maneuverable.

b. Antitank grenade.—Antitank grenades (M9) are fired from an M1903 rifle. These explosive charges have a short range (75 yards), but they do terrific damage to a tank. They are easily transported, and individual soldiers can destroy a tank with them.

c. Antitank rocket.—The antitank rocket is a new weapon—our Army's destructive answer to the tank. In the hands of the soldier, it is a powerful tank-destroying instrument, accurate up to 300 yards.

d. Frangible grenades ("Molotov cocktails").—These are incendiary grenades or improvised bottled inflammable liquid mixed with sawdust. A number of them thrown at the upper part of a tank and ignited will set the tank on fire.

e. Mines.—Antitank mines stop a tank and allow antitank fire to be brought upon it.

■ 59. THE HUNT.—Tank hunting follows a simple pattern, varied according to whether it is day or night.

a. Daytime hunt.—The daytime technique consists of setting a trap in a tank defile. A tank defile is a route which forces a tank to adhere to a certain path: for example, a road cut into a hill and surrounded by steep cliffs, or a road passing through an otherwise impassable bog. By means of mines the tank is confined in a limited space, and the hunters destroy it with their weapons. Smoke is used to conceal the activity of the hunters once the tank is trapped.

b. Nighttime hunt.—At night tank crews rest themselves and their tanks. The usual procedure is for the crews to get out of the tanks and rest in the immediate vicinity, posting sentinels to guard against attack. By means of stealth, such tank bivouacs can be attacked successfully. Part of the attacking party is detailed to take care of the crew members, while another party is assigned to destroy the tanks with hand-placed charges.

SECTION II
ANTITANK DEFENSE

■ 60. SECURITY.—Antitank defense is based upon two objectives: to prevent surprise and to stop tanks, by means of obstacles, long enough to destroy them with antitank fire. To accomplish the first of these objectives a constant system of sentinels is maintained; to accomplish the second a well-integrated system of obstacles, always defended by antitank and small-arms fire, is used. The small-arms fire prevents the removal of the obstacles by enemy engineers.

Engineers recover an M4 Sherman that had been disabled by German anti-armor mines in 1944 in Italy. (Signal Corps Archive)

■ 61. OBSTACLES.—*a. Mine fields.*—Mine fields are laid in definite patterns, two of which are shown in figure 108. Patterns are designed to do two things:

(1) Stop the tank before it gets through the field.

(2) Make it easier for our own engineers to remove the field later on. The details of mine-field laying are taken up in FM 5-30.

b. Ditches.—Antitank ditches have sides steep enough to stop a tank.

c. Posts and logs.—Post and log obstacles are effective in stopping or "bellying up" a tank.

d. Abatis.—A road block made with fallen trees can stop a tank for a time, if the trees are big enough.

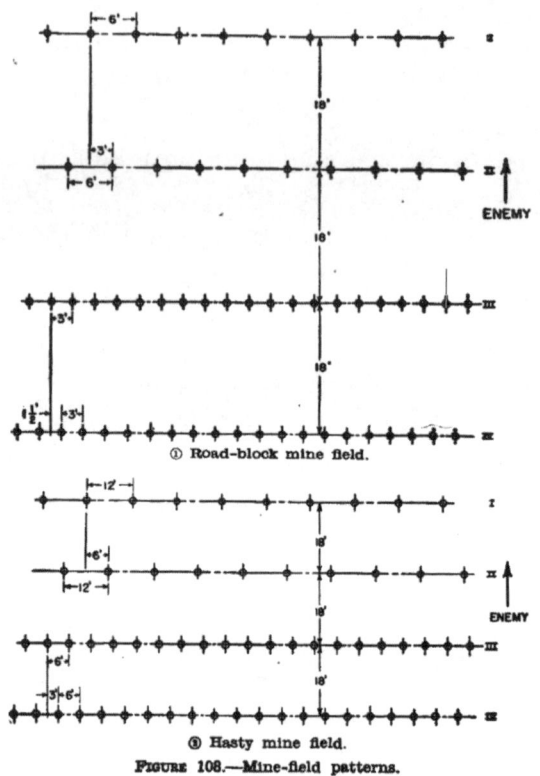

① Road-block mine field.

⑧ Hasty mine field.

FIGURE 108.—Mine-field patterns.

CHAPTER 8
ASSAULT OF A FORTIFIED POSITION

■ 62. THE PROBLEM.—a. A typical prepared defensive system of fortifications consists of a number of mutually supporting strong points, such as concrete emplacements called "pill boxes." The best way many of these can be destroyed is by foot troops armed with special weapons. It's a difficult combined-arms job to which engineers are often assigned. It requires aggressiveness, skill, speed, teamwork, courage, and determination. This chapter outlines procedure for a simple assault on a single fortified emplacement containing men and guns which fire from loopholes or embrasures. However, it must be remembered that pillboxes in an area are sited for mutual support; the whole problem is more complicated than this one.

b. The fortification is in a strong position. It is well placed; its walls resist bombardment; it generally has an open area around it so that its guns can cover a lot of ground. But, as the attacker, you have a number of important advantages:

(1) You are free to move around in the area, while the emplacement is stationary.

(2) The emplacement has blind spots, especially once you are close to it. It can fire only out of its loopholes.

(3) Once you get near it, the emplacement can't fire at you.

■ 63. PREPARATION.—Much training is required to assault a fortified position, and the teamwork is carefully planned. Each individual in the attacking force has a definite job to do at a certain time. He must accomplish his task or the efforts of the whole force may fail.

■ 64. ORGANIZATION.—A typical assault echelon for the attack of an emplacement is composed of two platoons—an assault platoon and an infantry rifle platoon. The infantry platoon attacks and neutralizes the earthen entrenchments and emplacements which are near the fortified emplacement and which cover the fortified emplacement with their fire. The assault platoon, which may include engineers, has two sections: the assault detachment and the support. It is this assault detachment which finally reaches and destroys the pillbox.

■ 65. ATTACK.—The attack proceeds, generally, in the following steps:

a. Artillery and airplanes bombard emplacement.

b. Direct-fire weapons fire at embrasures.

c. A special detachment breaches bands of obstacles to prepare way for assault echelon.

d. Assault echelon attacks.

■ 66. ASSAULT PLATOON.—*a.* The assault platoon works on a simple plan: one part of the assault platoon "covers" the advance of the second part until the fort is reached and the guns can be silenced by hand-placed charges. The covering section may consist of men armed with "tommy guns," machine guns, pistols, grenades, and rifles, which are aimed at the gun slits in the emplacement in order to stop the fire of the defenders. The advancing group moves forward in bounds, taking advantage of shell holes and other cover. Smoke is used to cover the advance.

b. The forward element of the assault detachment has two main parts: flamethrowers, who get close to the pillbox and blind its occupants with fire and smoke; and charge placers, who rush to the fortification and thrust into the weak spots of the fort (doors and embrasure openings) TNT attached to the end of long poles.

c. After one pillbox is silenced the assault group reorganizes and moves on to the next pillbox.

[. . .]

CHAPTER 14
COMBAT WEAPONS

■ 90. FIGHTING ENGINEERS.—The big job of engineers is construction and demolition in order to assist our movement and hinder that of the enemy. That job doesn't leave much spare time for fighting. However, the engineer is a scrapper and is given combat weapons with which to protect himself at work and so that he can reinforce the infantry when necessary.

■ 91. COMBAT WEAPONS.—The principal weapons of the combat engineer and their characteristics are as follows:

a. Hand grenades.—(1) Offensive grenades.—Depend upon blast effect only. No fragmentation. Effective bursting radius—5 yards. Should be used when thrower lacks cover to protect himself from flying fragments. Can be used for light demolitions and as priming charge for heavier demolitions.

(2) *Defensive grenades.*—Fragmentation type. Bursting radius—30 yards. Can cause casualties up to 200 yards. Should be thrown from covered positions, or into fox holes, trenches, or other enclosures, to prevent injuries to thrower. An excellent weapon against crew-served weapons in emplacements.

(3) *Smoke grenades, WP* [White Phosphorous] *or HC* [hexachloroethane, smoke].—Used to conceal your own activities, or to blind the enemy and hamper his fire and movement. HC has a slight irritant effect. WP can cause severe burns.

(4) *Frangible grenades.*—For antitank incendiary use. Consists of a glass bottle filled with gasoline, or other inflammable material with an igniter, which causes it to burst into flame when broken. Effective when thrown into open tank hatches or air intake ports.

(5) *Thermite grenades.*—For destruction of material. Emits white-hot molten metal that burns through light metal. Useful in igniting gasoline or oil in drums or other inflammable materials in metal containers.

b. Antitank rifle grenade discharger.—A short-range antitank rifle grenade, projected from a discharger fitted on a caliber .30 rifle. Maximum range against tanks is 75 yards. Penetrates any known light or medium tank. Penetration not influenced by range.

c. Antitank rocket discharger ("bazooka").—An armor-piercing weapon that breaches armor of any known light or medium tank. Much more powerful than antitank rifle grenade. Maximum effective range under favorable conditions is 300 yards, beyond which it is comparatively inaccurate. Penetration not affected by range. Primarily an antitank weapon, though it may be used effectively against crew-served weapons and point targets other than tanks.

d. Bayonet.—For shock action. All crack troops are good bayonet fighters.

e. Pistol, caliber .45.—A self-loading weapon carried by senior officers for close protection.

f. Submachine gun, caliber .45.—A short-range automatic weapon, excellent for close combat in an emergency situation.

g. Carbine, caliber .30.—A self-loading weapon, very effective up to 300 yards. An excellent medium-range rifle, very light and handy. Issued to company officers, key noncommissioned officers, officers, and messengers in combat battalions; is basic arm for rear area.

h. Rifle, caliber .30, M1.—A self-loading weapon, very effective up to 600 yards. The fundamental engineer combat weapon and the best of its type.

i. Machine gun, caliber .30, heavy.—A water-cooled automatic weapon, capable of a high rate of sustained fire. Used to provide base of fire in attack. Lays down final protective lines and covers sectors of fire in organized defenses. Excellent for covering mine fields and obstacles to prevent their removal. Covers approaches to bivouacs and working parties.

j. Machine gun, caliber .30, light.—Automatic air-cooled weapon, with a comparatively low rate of sustained fire.

k. Machine gun, caliber .50.—When suitably mounted, an effective anti-aircraft weapon, especially when rounds in belt contain mixture of tracer, armor-piercing, and incendiary bullets. Also excellent for knocking out trucks and lightly armored vehicles.

■ 92. GENERAL.—The following fundamentals should be remembered by every soldier—

a. Know your weapons and be able to hit with them.

b. Always take your weapons to work and keep them ready for use. Every job must have local security.

c. Shoot only when you have something to shoot at and are reasonably sure of hitting your target. It is a recruit trick to disclose a maneuver or position by firing too soon. This probably is the worst individual mistake in combat.

Sources

Introduction

War Department, *Engineer Field Manual, Volume I: Engineer Troops* (Washington, D.C., US Government Printing Office, 1932)

Chapter 1

Fowle, Barry E. (ed.), *Builders and Fighters: U.S. Army Engineers in World War II* (Fort Belvoir, VA, Office of History USACE, 1992)

War Department, *Intelligence Bulletin* (Washington, D.C., US Government Printing Office, October 1942)

War Department, FM 5-5, *Engineer Field Manual: Engineer Troops* (Washington, D.C., US Government Printing Office, 1943)

Chapter 2

War Department, FM 21-105, *Basic Field Manual: Engineer Soldier's Handbook* (Washington, D.C., US Government Printing Office, 1943)

War Department, FM 5-5, *Engineer Field Manual: Engineer Troops* (Washington, D.C., US Government Printing Office, 1943)

War Department, FM 5-10, *Engineer Field Manual: Communications, Construction, and Utilities* (Washington, D.C., US Government Printing Office, 1940)

War Department, TM-E 30-480, *Handbook on Japanese Military Forces* (Washington, D.C., US Government Printing Office, October 1944)

Chapter 3

War Department, FM 5-10, *Engineer Field Manual: Communications, Construction, and Utilities* (Washington, D.C., US Government Printing Office, 1940)

War Department, *Combat Lessons*, No. 7 (Washington, D.C., US Government Printing Office, 1944)

War Department, FM 21-105, *Basic Field Manual: Engineer Soldier's Handbook* (Washington, D.C., US Government Printing Office, 1943)

Chapter 4

War Department FM 5-15, *Corps of Engineers: Field Fortifications* (Washington, D.C., US Government Printing Office, 1944)

War Department, *Intelligence Bulletin* (Washington, D.C., US Government Printing Office, August 1943)

Chapter 5

War Department, FM 21-105, *Basic Field Manual: Engineer Soldier's Handbook* (Washington, D.C., US Government Printing Office, 1943)

War Department, *Combat Lessons*, No. 6 (Washington, D.C., US Government Printing Office, 1944)

War Department, Pamphlet No. 21–23, *Don't get killed by Mines and Booby Traps* (Washington, D.C., US Government Printing Office, 1944)

Chapter 6

War Department, FM 5-6, *Engineer Field Manual: Operations of Engineer Field Units* (Washington, D.C., US Government Printing Office, 1943)

HQ 28th Infantry Division, *After-Action Report*, APO 28, U.S. Army (January 4, 1945)

War Department, FM 21-105, *Basic Field Manual: Engineer Soldier's Handbook* (Washington, D.C., US Government Printing Office, 1943)